LIVERPOOL JMU LIBRARY

Methods in Microbiology
Volume 40

Recent titles in the series

Volume 24 *Techniques for the Study of Mycorrhiza*
JR Norris, DJ Reed and AK Varma

Volume 25 *Immunology of Infection*
SHE Kaufmann and D Kabelitz

Volume 26 *Yeast Gene Analysis*
AJP Brown and MF Tuite

Volume 27 *Bacterial Pathogenesis*
P Williams, J Ketley and GPC Salmond

Volume 28 *Automation*
AG Craig and JD Hoheisel

Volume 29 *Genetic Methods for Diverse Prokaryotes*
MCM Smith and RE Sockett

Volume 30 *Marine Microbiology*
JH Paul

Volume 31 *Molecular Cellular Microbiology*
P Sansonetti and A Zychlinsky

Volume 32 *Immunology of Infection, 2nd edition*
SHE Kaufmann and D Kabelitz

Volume 33 *Functional Microbial Genomics*
B Wren and N Dorrell

Volume 34 *Microbial Imaging*
T Savidge and C Pothoulakis

Volume 35 *Extremophiles*
FA Rainey and A Oren

Volume 36 *Yeast Gene Analysis, 2nd edition*
I Stansfield and MJR Stark

Volume 37 *Immunology of Infection*
D Kabelitz and SHE Kaufmann

Volume 38 *Taxonomy of Prokaryotes*
Fred Rainey and Aharon Oren

Volume 39 *Systems Biology of Bacteria*
Colin Harwood and Anil Wipat

Methods in Microbiology
Volume 40

Microbial Synthetic Biology

Edited by

Colin Harwood
Institute of Cell and Molecular Biosciences
Baddiley-Clark Building
Newcastle University
Newcastle upon Tyne
NE2 4AX

Anil Wipat
School of Computing Science
Claremont Tower
Newcastle University
Newcastle upon Tyne
NE1 7RU

AMSTERDAM • BOSTON • HEIDELBERG • LONDON
NEW YORK • OXFORD • PARIS • SAN DIEGO
SAN FRANCISCO • SINGAPORE • SYDNEY • TOKYO
Academic Press is an imprint of Elsevier

Academic Press is an imprint of Elsevier
The Boulevard, Langford Lane, Kidlington, Oxford, OX5 1GB, UK
32 Jamestown Road, London NW1 7BY, UK
Radarweg 29, PO Box 211, 1000 AE Amsterdam, The Netherlands
225 Wyman Street, Waltham, MA 02451, USA
525 B Street, Suite 1800, San Diego, CA 92101-4495, USA

First edition 2013

Copyright © 2013 Elsevier Ltd. All rights reserved.

No part of this publication may be reproduced, stored in a retrieval system or transmitted in any form or by any means electronic, mechanical, photocopying, recording or otherwise without the prior written permission of the publisher.

Permissions may be sought directly from Elsevier's Science & Technology Rights Department in Oxford, UK: phone (+44) (0) 1865 843830; fax (+44) (0) 1865 853333; email: permissions@elsevier.com. Alternatively you can submit your request online by visiting the Elsevier web site at http://elsevier.com/locate/permissions, and selecting Obtaining permission to use Elsevier material.

Notice
No responsibility is assumed by the publisher for any injury and/or damage to persons or property as a matter of products liability, negligence or otherwise, or from any use or operation of any methods, products, instructions or ideas contained in the material herein. Because of rapid advances in the medical sciences, in particular, independent verification of diagnoses and drug dosages should be made.

> For information on all Academic Press publications
> visit our website at www.store.elsevier.com

ISBN: 978-0-12-417029-2
ISSN: 0580-9517 (Series)

Printed and bound in UK

13 14   10 9 8 7 6 5 4 3 2 1

Contents

Contributors xi
Preface xiii

CHAPTER 1 **Computational Intelligence in the Design of Synthetic Microbial Genetic Systems** 1
Jennifer S. Hallinan
1. Introduction 1
2. Computational Infrastructure for Synthetic Biology 2
   2.1. Biological Parts 2
   2.2. Circuit Design and Simulation 3
   2.3. Exploring Design Space 8
3. Computational Intelligence 9
   3.1. Evolutionary Algorithms 10
   3.2. Other CI Techniques 23
4. Discussion 27
   References 29

CHAPTER 2 **Constraints in the Design of the Synthetic Bacterial Chassis** 39
Antoine Danchin, Agnieszka Sekowska
1. Introduction 39
2. Top-Down Versus Bottom-Up Framing 40
3. An Unlimited List of Functions (Organised Starting from the Cell's Structure) 43
   3.1. Compartmentalisation and Shaping: the Cell's Casing 43
   3.2. Information Transfer 47
   3.3. Metabolism 55
4. Perspective: The Fourth Dimension (When Time Measures and Shapes) 57
   Acknowledgements 59
   References 59

CHAPTER 3 **Social Dimensions of Microbial Synthetic Biology** 69
Jane Calvert, Emma Frow
1. Introduction: Making Space for a New Discussion 69
2. From Implications to Dimensions 70
   2.1. Metaphors and Analogies 71
3. From Speculation to Anticipation 72

v

4. From Public Acceptance to Public Good 75
   4.1. Synthetic Biology for a Particular Purpose—the
        Arsenic Biosensor 77
   4.2. The Public Good 79
5. From Regulation to Governance 80
   5.1. Responsible Research and Innovation 81
6. Conclusions 82
   References 83

CHAPTER 4 *Bacillus subtilis*: Model Gram-Positive Synthetic Biology Chassis 87
Colin R. Harwood, Susanne Pohl, Wendy Smith, Anil Wipat

1. Introduction 87
2. The *B. subtilis* Genome 88
   2.1. General Features 88
   2.2. Genome Annotation 89
3. Genome Management and Analysis of Gene
   Function 89
   3.1. Gene Transfer and Recombination 89
   3.2. Transformation 90
   3.3. Conjugation 91
   3.4. Plasmid-Based Vector Systems 91
   3.5. Special-Purpose Vectors 92
   3.6. Expression Vectors 101
   3.7. Genome Minimalisation 102
4. Analysis of the Transcriptome 102
   4.1. Transcription and Transcription Profiling 102
   4.2. Sigma Factors 103
   4.3. Transcription Termination 104
   4.4. Reporter Gene Technology 104
   4.5. Reporter Gene Libraries/Live Cell Arrays 104
   4.6. Analysis of Populations and Single Cells 105
5. Analysis of the Proteome 105
6. Analysis of the Metabolome 106
7. Parts, Devices, Systems and Applications 107
   7.1. Parts, Systems and Devices 107
   7.2. Engineering Genomes with *B. Subtilis* 108
   7.3. Biosensors 108
8. Computational Tools and Resources 109

9. Summary 111
References 111

CHAPTER 5 Engineering Microbial Biosensors 119
Lisa Goers, Nicolas Kylilis, Marios Tomazou,
Ke Yan Wen, Paul Freemont, Karen Polizzi
1. Introduction 119
2. Areas of Application 120
   2.1. What to Target? 121
   2.2. Finding Potential Sensors 122
3. Types of Biosensors 122
   3.1. Response Characteristics 124
4. Reporters 128
   4.1. Fluorescence 128
   4.2. Bioluminescence 128
   4.3. Colour Change 128
5. Biosensors and Synthetic Biology 129
6. Transcription-Based Biosensors 129
   6.1. Planning the Construct 129
   6.2. Assembling the Construct 131
   6.3. Testing the Construct 132
7. Translation-Based Biosensors 134
   7.1. Planning the Construct 135
   7.2. Riboswitch Construction Guidelines 136
8. Posttranslational Biosensors 139
   8.1. Planning the Construct 139
   8.2. Assembling the Construct 140
   8.3. Testing the Construct 141
9. *In Vitro* Biosensors 141
10. Modelling Biosensors 143
    10.1. What is a Model and How to Plan the
          Modelling Process 143
    10.2. Constructing the Model 145
    10.3. Model Reduction 147
    10.4. Identifying the Starting Conditions and
          Parameter Values 148
    10.5. Making Use of the Model—Numerical Simulations 149
    10.6. Validating the Model and Reiterating the Modelling
          Procedure 150
11. Outlook 152
    References 153

CHAPTER 6 Noise and Stochasticity in Gene Expression: A Pathogenic Fate Determinant 157
Mikkel Girke Jørgensen, Renske van Raaphorst, Jan-Willem Veening

1. Introduction 157
2. Origins of Noise 158
   2.1. Transcriptional Bursting 158
3. Measuring Noise 160
4. Engineering Noise 162
5. Noise and Heterogeneity in Gene Expression 164
6. Bistable Expression of Pneumococcal Pili 165
7. Cooperative Virulence in *Salmonella enterica* S. Typhimurium 168
8. Concluding Remarks 170
   References 171

CHAPTER 7 Platforms for Genetic Design Automation 177
Chris J. Myers

1. Introduction 177
2. Standards 179
   2.1. Systems Biology Markup Language 179
   2.2. Synthetic Biology Open Language 180
3. Repositories 181
   3.1. BioModels Database 182
   3.2. GenBank 182
   3.3. Registry of Standard Biological Parts (iGEM Registry) 183
   3.4. Standard Biological Parts Knowledgebase 184
   3.5. BioFab 184
   3.6. BacilloBricks 186
   3.7. Inventory of Composable Elements 187
4. GDA Software Tools 187
   4.1. BioJADE 188
   4.2. GenoCAD 188
   4.3. TinkerCell 189
   4.4. Process Modelling Tool 190
   4.5. Synthetic Biology Software Suite 190
   4.6. Synthetic Biology Reusable Optimization Methodology 190
   4.7. Genetic Engineering of Cells 192
   4.8. Kera 194

      4.9. Intelligent Biological Simulator 194
      4.10. Sequence Editors and Optimizers 195
      4.11. Tool Chains 197
5. Discussion 198
      Acknowledgements 200
      References 200

Index **203**

Contributors

Jane Calvert
Science, Technology and Innovation Studies, University of Edinburgh, Old Surgeons' Hall, High School Yards, Edinburgh, United Kingdom

Antoine Danchin
AMAbiotics, Evry, France; Department of Biochemistry, Faculty of Medicine, The University of Hong Kong, Hong Kong, and CEA/Genoscope, Evry, France

Paul Freemont
Department of Life Sciences, and Centre for Synthetic Biology and Innovation, Imperial College London, London, United Kingdom

Emma Frow
Science, Technology and Innovation Studies, University of Edinburgh, Old Surgeons' Hall, High School Yards, Edinburgh, United Kingdom

Lisa Goers
Department of Life Sciences, and Centre for Synthetic Biology and Innovation, Imperial College London, London, United Kingdom

Jennifer S. Hallinan
School of Computing Science, Newcastle University, Newcastle upon Tyne, United Kingdom

Colin R. Harwood
Centre for Synthetic Biology and Bioexploitation, Newcastle University, Newcastle upon Tyne, United Kingdom

Mikkel Girke Jørgensen
Molecular Genetics Group, Groningen Biomolecular Sciences and Biotechnology Institute, Centre for Synthetic Biology, University of Groningen, Groningen, The Netherlands

Nicolas Kylilis
Department of Life Sciences, and Centre for Synthetic Biology and Innovation, Imperial College London, London, United Kingdom

Chris J. Myers
Department of Electrical and Computer Engineering University of Utah, Salt Lake City, Utah, USA

Susanne Pohl
Centre for Synthetic Biology and Bioexploitation, Newcastle University, Newcastle upon Tyne, United Kingdom

Karen Polizzi
Department of Life Sciences, and Centre for Synthetic Biology and Innovation, Imperial College London, London, United Kingdom

Agnieszka Sekowska
AMAbiotics, Evry, France; Department of Biochemistry, Faculty of Medicine, The University of Hong Kong, Hong Kong, and CEA/Genoscope, Evry, France

Wendy Smith
Centre for Synthetic Biology and Bioexploitation, Newcastle University, Newcastle upon Tyne, United Kingdom

Marios Tomazou
Department of Life Sciences; Centre for Synthetic Biology and Innovation, and Department of Bioengineering, Imperial College London, London, United Kingdom

Renske van Raaphorst
Molecular Genetics Group, Groningen Biomolecular Sciences and Biotechnology Institute, Centre for Synthetic Biology, University of Groningen, Groningen, The Netherlands

Jan-Willem Veening
Molecular Genetics Group, Groningen Biomolecular Sciences and Biotechnology Institute, Centre for Synthetic Biology, University of Groningen, Groningen, The Netherlands

Anil Wipat
Centre for Synthetic Biology and Bioexploitation, Newcastle University, Newcastle upon Tyne, United Kingdom

Ke Yan Wen
Department of Life Sciences, and Centre for Synthetic Biology and Innovation, Imperial College London, London, United Kingdom

Preface

Microbial synthetic biology is a rapidly growing discipline that builds on well-established principles of genetic engineering and biotechnology by integrating computational and engineering approaches into the design and construction of novel biological systems. Although synthetic biology represents a multidisciplinary technology with enormous potential to benefit society, major challenges need to be overcome to achieve a radical step change in our ability to engineer complex multi-scaled biological systems. At the heart of these challenges is a need to understand the enormous complexity that underlies even the simplest of biological entities. This requires a detailed understanding of cell physiology, biochemistry and genetics, as well as that of the sub-cellular molecular machinery that mediates essential biological processes.

The step change in technology that has brought synthetic biology to the fore is our ability to analyse very large datasets generated by high-throughput omics technology. The computational tools and approaches needed for the Computer-Aided Design (CAD) and Computer-Aided Manufacturing (CAM) of complex multi-scaled biological systems (and their constituent molecular machines), although still in their infancy, are already beginning to show their value. The chapters in this volume have been chosen to address the methodology associated with different aspects of synthetic biology.

Responsible innovation must be at the heart of the bioexploitation of biological systems. Calvert and Flow address the issue of responsible innovation and, in particular, the ethical, legal and social implications of synthetic biology. They challenge some of the more familiar ways of addressing these issues by proposing a series of shifts in the framing and language used to discuss the social aspects of synthetic biology.

The Chapter by Hallinan deals with computational intelligence and the design of synthetic microbial systems. It addresses the issue of genome-scale circuit design, one of the ultimate aims of synthetic biology. Critical to the success of synthetic biology is the development of platforms for the automation of genetic design. Myers outlines the standards needed for data representation that facilitates genetic design tasks and discusses the future requirements for genome-scale automated genetic design workflows.

Harwood and colleagues focuses on one of the most widely used synthetic biology chassis, *Bacillus subtilis*. This bacterium is already used extensively for the commercial production of proteins, enzymes and metabolites such as riboflavin. This chapter discusses both the techniques used for genome refactoring and the computational tools and resources required to model its metabolism. In their chapter, Danchin and Sekowska address the constraints on the design of synthetic bacterial chassis, contrasting the top-down and bottom-up approaches to chassis development. They also remind the reader of the importance of time, the fourth dimension of any three-dimensional biological system.

Biological systems are noisy, and individual genetically identical cells in a population subjected to the same environmental conditions exhibit a variety of phenotypes. Girke and colleagues explore the origins, measurement and engineering of noise and its influence on gene expression. As an example of the exploitation of synthetic biology, Goers and colleagues discuss the design, engineering and application of biosensors. Various types of biosensors are described together a range of reporters used to generate an observable signal. This chapter describes how mathematical models can be used to predict the characteristics and performance of biosensors and aid their design.

We sincerely hope that this volume provides the reader with valuable insights into the technology and methods currently being developed for the rapidly expanding discipline of synthetic biology.

Colin Harwood and Anil Wipat

CHAPTER 1

Computational Intelligence in the Design of Synthetic Microbial Genetic Systems

Jennifer S. Hallinan[1]

School of Computing Science, Newcastle University, Newcastle upon Tyne, United Kingdom
[1]*Corresponding author. e-mail address: jennifer.hallinan@ncl.ac.uk*

1 INTRODUCTION

The oft-quoted defining characteristic of synthetic biology is "the application of engineering principles to biological systems" (Andrianantoandro, Basu, Karig, & Weiss, 2006). This approach is fundamental to the discipline. Engineering concepts such as abstraction, modularity and the use of standard interfaces are clearly valuable for any design enterprise. On a small scale, engineering approaches can be applied manually to individual designs. The true promise of synthetic biology will, however, require biological engineering on a much larger scale. It is becoming apparent that the original aim of synthetic biology—the reduction of biological parts and devices to the equivalent of children's building blocks (Heinemann & Panke, 2006)—is optimistically oversimplified; biological systems are messy, stochastic and highly interconnected and do not lend themselves easily to division into tidy components (Agapakis, Boyle, & Silver, 2012).

However, in order to rationally design microbial cells with predictable, desirable functionality, some level of abstraction away from this complexity must be achieved. Small circuits can be designed using expert knowledge applied to the manipulation of tiny, abstract modules such as BioBricks (Shetty, Endy, & Knight, 2008), as demonstrated by hundreds of successful iGEM projects, and a plethora of 'proof of principle' circuits implementing logic gates, inverters, oscillators and their ilk. Practically useful and commercially viable synthetic biology will, however, necessarily involve much larger-scale designs.

Large-scale—even genome-scale—circuit design is often presented as an ultimate aim of synthetic biology (Barrett, Kim, Kim, Palsson, & Lee, 2006; Carrera, Rodrigo, & Jaramillo, 2009; Searls, Guan, Dunham, Caudy, & Troyanskaya, 2010; Esvelt & Wang, 2013). The reasons for this desire are twofold: for a start, the ability to create an entire organism, no matter how crippled, will provide us with a much more complete understanding of the systems-level organisation of bacteria than we currently possess. Additionally, small-scale designs are inherently limited in the behaviours that they can produce, while larger-scale designs are theoretically able to generate practically any biologically plausible behaviour envisaged by their designers. Some progress towards this aim has been made by groups such as those led by Craig Venter (Gibson et al., 2010) and George Church (Forster & Church, 2006; Carr et al., 2012), but there is still a long way to go before this task is routine (Noireaux, Maeda, & Libchaber, 2011).

Large-scale designs will have to take into account practical issues such as optimal codon usage, interactions between microbes and their environments and crosstalk between existing and novel pathways within an organism. In order to maximise the use of existing data and to identify, combine and evaluate biologically meaningful abstractions of parts, devices, pathways and even genomes, productive, large-scale synthetic biology will require computational data integration, modelling, simulation and analysis.

2 COMPUTATIONAL INFRASTRUCTURE FOR SYNTHETIC BIOLOGY

Computer science can contribute to synthetic biology in numerous ways, from the establishment of community-based standards (Smith et al., 2007), through the development of freely available Web-based resources (Fernández-Suárez & Galperin, 2013), to the building of automated eScience workflows (Craddock, Harwood, Hallinan, & Wipat, 2008). Perhaps the most immediately valuable contributions, however, are the application of automated data integration to the identification and parameterisation of biological parts; automated circuit design; the simulation of models of novel circuits; measurement of the behaviour of engineered circuits, both *in silico* and *in vivo*; and automated computational reasoning for the analysis of existing behaviour and the generation of testable hypotheses leading to modification of models or the design of new experiments.

2.1 Biological parts

Conceptually, most synthetic biologists think in terms of *parts* (basic sequence components such as promoters and coding sequences) and *devices* (collections of parts that work together to produce a desired behaviour, such as a sensor or bistable switch) (Endy, 2005; Canton, Labno, & Endy, 2008). In practice, the construction of synthetic genetic circuits has traditionally, in so far as a field as young as synthetic biology may be said to have traditions, been carried out using sequence-based parts such as Bio-Bricks. BioBricks are short DNA sequences with defined 5' and 3' ends, designed to enable rapid combination of different BioBricks using conventional cloning, without leaving a scar on the sequence. Each BioBrick has a clearly defined function, which may be as simple as detecting a specific input molecule or as complex as sensing multiple inputs and producing any of a number of desired outputs. The major repository of BioBricks is the Registry of Standard Biological Parts (Canton et al., 2008).[1]

---

[1] http://partsregistry.org/Main_Page

Synthetic biologists are not, however, restricted to designs using combinations of small parts. Technologies for DNA synthesis and sequencing are developing rapidly. The first gene synthesised, in 1965, was for an alanine–tRNA. Assembly of its 77 nucleotides took 5 years (Czar, Anderson, Bader, & Peccoud, 2009). At the time of writing, DNA synthesis is commercially available from around $US0.28 per base pair, and up to 7.5 million base pairs can be synthesised per month.[2] The advent of fast and relatively cheap synthesis means that the usefulness of sequence-based approaches such as BioBricks is becoming somewhat moot. Novel circuits can be designed *in toto* and synthesised relatively quickly and cheaply. Rapid DNA synthesis is convenient for the researcher and has several other advantages over conventional cloning for synthetic biology.

A major aim of many synthetic biology projects is the production or overproduction of specific proteins or metabolites. The ability to combine protein-coding sequences from a variety of organisms into a single circuit, or to design entirely novel proteins, gives the synthetic biologist considerable flexibility in the design of novel systems, but there are some caveats. Different organisms have different codon usage biases, meaning that proteins from one organism can be difficult to express in a heterologous host (Gustafsson, Govindarajan, & Minshull, 2004). Codon optimisation can lead to significant increases in protein expression levels.

The use of large-scale DNA synthesis, rather than conventional cloning, means that sequence-level modifications such as codon usage optimisation can be designed *in silico* and implemented efficiently. In addition, modifications to a basic design can be produced quickly and cheaply. DNA synthesis companies keep copies of an original design, and minor sequence modifications (which may have major phenotypic effects) can be produced at considerably less cost than that incurred by a completely new sequence.

Whether by combining BioBricks in the lab or by sending digital sequence files to a synthesis company, construction of novel synthetic genetic circuits is currently a relatively straightforward, inexpensive process. However, before a circuit can be constructed, it must be designed and the design evaluated.

2.2 Circuit design and simulation
2.2.1 Bottom-up versus top-down design
The design approach currently dominant in synthetic biology is essentially manual. Once a problem is identified, and any necessary requirements established, a domain expert examines pathways known or suspected to be involved in the biological function of interest. Potentially useful modifications to existing pathways are identified, and a programme of laboratory experimentation is devised. This approach is designated *bottom-up* design, because it starts with a consideration of the basic components—parts and devices—which may be used and builds the systems up from there (Andrianantoandro et al., 2006; Schwille, 2011).

---

[2] http://www.genscript.com/gene_synthesis.html

This bottom-up design approach has a number of advantages. Perhaps most importantly, it is intuitive and readily comprehensible to those with the appropriate background knowledge. The logic and reasoning behind a given circuit design can be explained in detail. The behaviour of organisms carrying pathways designed in a bottom-up manner may be relatively straightforward to interpret, since the behaviour of the individual components is relatively well understood.

However, the bottom-up approach also has a number of significant problems. Most obvious is the need for detailed information, not only about individual parts and their behaviour but also about the ways in which they interact. Even for relatively well-studied model organisms, such information is simply not available for a significant proportion of genes. At the time of writing, the Genoscope database entry (Barbe et al., 2009)[3] for the important model Gram-positive bacterium *Bacillus subtilis* lists 850 (16%) of its 4533 CDS as having 'unknown' or 'hypothetical' function. Interactions involving these proteins are also, perforce, uncharacterised, as are the majority of other biophysical interactions, even between relatively well-studied entities. The situation becomes even more difficult when one considers the numerous RNAs and subtly different proteins to which posttranscriptional modifications may give rise.

The number of potential interactions between a set of parts increases exponentially with the size of the set. Consider a simple circuit consisting of a promoter, a ribosomal binding site (RBS), a coding sequence (CDS) and a terminator. If there are two possible promoters, two RBSs and two terminators available (assuming that the CDS is an immutable readout), the possible number of circuits is $2 \times 2 \times 1 \times 2 = 8$. For three choices of each part, the number of combinations is 27; the number of possible combinations increases exponentially with the number of parts available. In the real world, a number of groups have investigated the generation and characterization of libraries of promoters (De Mey, Maertens, Lequeux, Soetaert, & Vandamme, 2007; Hartner et al., 2008; Blount, Weenink, Vasylechko, & Ellis, 2012), RBSs (Dougherty & Arnold, 2009; Chen & Arkin, 2012) and terminators (Temme, Hill, Segall-Shapiro, Moser, & Voigt, 2012). Such libraries can consist of tens or even hundreds of individual components, most of which are only partially characterised. Predicting the behaviour arising from the interactions between these components in all of their potential combinations is simply not possible for an individual microbiologist, no matter how expert. Bottom-up designs, with their dependence upon expert knowledge, are therefore necessarily limited in size.

At a more fundamental level, bottom-up designs are restricted to those that can be conceived of, and implemented, by humans. The number of viable, productive biological pathways produced by the evolutionary process is a tiny fraction of those that potentially exist in the vast solution space defined by four bases, 20 amino acids and millions of possible interactions between biomolecules. Given that the evolutionary interests of naturally occurring microbes are not necessarily the same as those of

---

[3]https://www.genoscope.cns.fr/agc/microscope/mage/index.php?

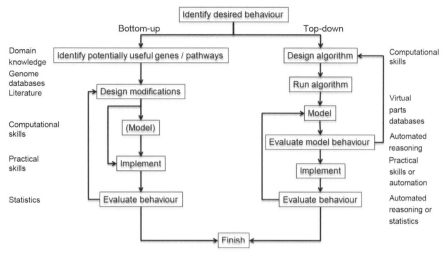

FIGURE 1.1
Bottom-up versus top-down design of synthetic circuits. The steps of the design process are indicated in rectangles, the resources required are listed alongside.

human synthetic biologists, it is highly unlikely that optimal designs for genetic circuits performing totally novel functions will be found solely amongst those already known to researchers. Bottom-up design restricts synthetic biology in both scale and scope.

The alternative to bottom-up design is, unsurprisingly, known as *top-down* design (O'Malley, Powell, Davies, & Calvert, 2008; Figure 1.1). With top-down design, the desired phenotype of a microbe is specified, and the actual design work handed off to an algorithm. An algorithm is "a process or set of rules to be followed in calculations or other problem-solving operations, especially by a computer".[4] A top-down algorithm takes as input two pieces of information: a desired phenotype and a set of parts that can be combined to produce this phenotype. Since the algorithm is computational, the description of the parts must be in a format that can be understood and manipulated by computers.

2.2.2 Computer-aided design
Computer-aided design (CAD) is ubiquitous in the design of electrical and electronic circuits. It is widely acknowledged that complex electrical circuits cannot be designed by the human mind alone (Chandrakasan, Bowhill, & Fox, 2000). Large-scale systems in domains as complex as living organisms, even those as amenable as microorganisms, are also clearly impossible to design manually, and it is

---

[4]http://oxforddictionaries.com/definition/english/algorithm

becoming increasingly appreciated that CAD systems will be essential for large-scale synthetic biology (Heinemann & Panke, 2006).

Computational approaches can, of course, be applied to every aspect of the synthetic biology life cycle. For recent overviews of the entire computational design process as applied to synthetic biology, see Marchisio and Stelling (2009) and Medema, van Raaphorst, Takano, and Breitlingg (2012). A list of tools covering every aspect of computational design in synthetic biology is available at http://openwetware.org/wiki/Computational_Tools. In this chapter, we cover only the application of computational approaches to the *design* of synthetic circuits. There are numerous CAD systems designed specifically for synthetic biology, with varying levels of maturity.

In order to design genetic circuits using computational approaches, biological parts must be represented in a form that can be manipulated by computers; that is, they must be represented as *virtual parts*.

2.2.3 Standard virtual parts

There are a number of different, computationally amenable formats describing biological parts and devices. Probably the most well known is the BioBricks standard. BioBricks are physical DNA sequences meeting specific conditions for the sequence of 5′ and 3′ ends, presence or absence of restriction sites and so forth.[5] The Registry of Standard Biological Parts also contains metadata about the BioBricks and is computationally accessible, via SOAP Web services (Curbera et al., 2002). Several CAD tools have been written specifically to manipulate BioBricks, all of which operate using a bottom-up approach (Chandran, Bergmann, & Sauro, 2009; Weeding, Houle, & Kaznessis, 2010; Chen, Densmore, Ham, Keasling, & Hillson, 2012).

Many researchers use libraries of parts that, while not following the BioBricks format, are similarly sequence-oriented (Suarez, Rodrigo, Carrera, & Jaramillo, 2010). Many of the CAD tools designed for BioBricks will also handle alternative parts representations. Such libraries are well suited to bottom-up device design, but the emphasis upon sequence means that the user must supply much of the required essential information. Basic knowledge such as the nature and function of components exists only in the users' brains, while more subtle issues, such as the optimum modifications to make to an existing circuit in order to produce a desired behaviour, are purely subjective and depend upon human expertise. The bottom-up design of synthetic circuits, using sequence-based parts, is a time- and knowledge-intensive task and does not lend itself well to automation or the exploration of alternative solutions to a given problem.

An alternative approach to the computational design of genetic circuits is to incorporate as much information as possible into the parts themselves. The nature of a parts-based model necessarily becomes more sophisticated, incorporating not only information about the sequence but also additional annotations regarding the nature of the part, relevant kinetic data and part provenance, such as the name of

---

[5]http://openwetware.org/wiki/The_BioBricks_Foundation:RFC

the designer, and the intended uses of the part. Such annotations must, of course, be in machine-readable format, to enable computational algorithms to access, interpret and manipulate the information (Misirli et al., 2011).

In a field as dynamic and diverse as synthetic biology, it is important wherever possible to use community-established standards, both for software and for annotation. There are a large number of standards addressing different aspects of biomedicine, existing under the umbrella of the Open Biological and Biomedical Ontologies (OBO) foundry.[6] OBO includes ontologies covering many different aspects of biomedicine, in various stages of development. Several of them are of potential value to synthetic biology; for example, the Sequence Ontology[7] specifies "a set of terms and relationships used to describe the features and attributes of biological sequence" (Eilbeck et al., 2005) and as such is useful for the standardised annotation of biological parts. The use of community-developed ontologies for parts annotation has the additional advantage that annotations can take the form of a hyperlink to a uniform resource identifier (URI), a string of characters used to identify a name or a Web resource. The amount of text used to describe a part within its text representation is thus minimised.

The second way in which basic sequence-based parts can be extended is by implementing them in a standard format, in a widely used modelling language that imposes its own constraints on the way in which parts may be combined. Several of the OBO standards, particularly those dealing with computational modelling and simulation, arise from the systems biology community. The systems biology modelling language (SBML) (Hucka et al., 2003) and CellML (Cooling, Hunter, & Crampin, 2008) are two widely used modelling languages. These languages support modularity of model design and the automated assembly of models from a 'bag' of parts. The input to one part can automatically be matched by the simulator to the output of another part.

One implementation of abstract parts models, suitable for automated top-down design, is standard virtual parts (SVPs) (Cooling et al., 2010; Misirli et al., 2011). SVPs consist of SBML models of biological parts, annotated with a specific, defined set of information, derived, wherever possible, from community-developed standards. SVPs can be automatically combined into simulateable models, forming the basis of an automated synthetic circuit design process.

The establishment of standards for parts models, be they BioBricks, SVPs or any other type of model, is only the start of the circuit design process. Individual parts must be combined into models of functional circuits with a desired behaviour, and ideally some idea of the efficiency and practicality of the circuit obtained, via simulation and computational analysis, before the circuit is synthesised, implemented into a host chassis and evaluated. In general, the results of the evaluation of *in vivo* behaviour will feedback into the systems design, establishing an iterative process of

---

[6] http://www.obofoundry.org/
[7] http://www.sequenceontology.org/

modelling, implementation and evaluation (Hallinan, Park, & Wipat, 2012; Figure 1.2). The results of this synthetic biology life cycle should be the identification of optimal, or near-optimal, circuit designs from amongst the huge number of possibilities that lurk in design space.

2.3 Exploring design space

The concept of 'design space' comes from physics and is a useful way in which to think about the possibilities inherent in the design process. Consider a simple genetic construct (Figure 1.3) that senses subtilin, a lantibiotic used by some strains of *Bacillus subtilis* for quorum communication (Klein, Kaletta, Schnell, & Entian, 1992). In response to subtilin, this circuit produces the green fluorescent protein (GFP), an easily detected reporter protein. Such a circuit could be the first stage of a larger design effort, in which another, more complex and sophisticated construct replaces the GFP

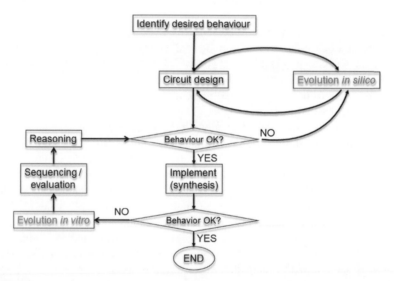

FIGURE 1.2

A dual evolutionary strategy for synthetic biology. (For colour version of this figure, the reader is referred to the online version of this chapter.)

FIGURE 1.3

A simple synthetic genetic circuit. (See color plate.)

response. The essential elements of the circuit are the genes for the bacterial two-component system that senses and responds to subtilin, the sensor kinase (SpaK) and the response regulator with which it interacts (SpaR), together with the gene-encoding GFP and appropriate promoters, operators and terminators.

The design depicted in Figure 1.3 is not the only possible construct that will perform the desired behaviour: that is, responding to subtilin with a GFP output. Different promoters, operators and terminators could be used, and the design could be made more complicated by the incorporation of additional interacting genes producing feedback loops, performing posttranscriptional modifications and so on. The complete set of all possible circuit designs that can produce the target behaviour is known as the *design space* for this problem (Thompson, Layzell, & Zebulum, 1999; Romer & Mattern, 2004). Even for a small circuit such as the one shown here, the design space is potentially very large (Edwards & Glass, 2000).

So-called *brute force* techniques for design space exploration involve constructing a model of each possible circuit, simulating its behaviour and evaluating the results. For complex designs, it quickly becomes computationally infeasible to explore the entire design space, no matter how sophisticated the available computational facilities. Design space almost certainly contains multiple, excellent solutions to any given problem. Many of these solutions may be counterintuitive, even to an experienced microbiologist, so it is important to explore design space in a principled, albeit necessarily heuristic, way.

The exploration of design space is facilitated by the fact that it has a degree of high-dimensional structure. It is easy to envisage a metric for closeness between two circuit designs. A circuit identical to the one shown in Figure 1.3, but with one of the promoters replaced by a different promoter, for example, is closer to the original than one in which multiple new components have been added. In design space, instances of a design are thus related both in terms of model structure and of *fitness*: how well a given circuit fulfils the design brief. Ideally, any approach to genetic circuit design should explore as much of the design space as possible, but do so in a principled manner. When a reasonably good design is identified, it makes sense to examine other designs that are structurally close to it, in the hopes that designs that are structurally related also have similar fitness. This tension between *exploration* of the design space and *exploitation* of already identified areas of high fitness is fundamental to space-searching and optimisation algorithms (Eiben & Schippers, 1998; Lin & Gen, 2009). Most of the algorithms that have been designed to search in this way fall into the broad category of 'computational intelligence'.

3 COMPUTATIONAL INTELLIGENCE

The field of computational intelligence (CI) comprises a range of algorithms designed to analyse datasets and draw generalisable conclusions from that data with a minimum of human intervention (Engelbrecht, 2007). CI approaches have been widely applied to problems in bioinformatics and biomedicine, resulting in the production, not only of a rich literature but also of a wide range of books (Fogel & Corne, 2003; Mitra, 2005; Fogel, Corne, & Pan, 2008; Kelemen, Abraham, & Chen, 2008), journals (such as the *International Journal of Computational Intelligence in Bioinformatics and Systems Biology*[8] and the *Journal of Computational Intelligence in Bioinformatics*[9]) and conferences (including the IEEE Symposium on Computational Intelligence in Bioinformatics and Computational Biology,[10] the AASRI Conference on Computational Intelligence and Bioinformatics[11] and the International Meeting on Computational Intelligence Methods for Bioinformatics and Biostatistics[12]). Because of the decade-long history of CI and biomedicine, the value of CI techniques for synthetic biology was recognised as soon as the field became mature enough to be considered as an independent area of research (Bongard, 2002).

CI algorithms fall into four broad categories: evolutionary algorithms (EAs), artificial neural networks (ANNs), fuzzy logic and swarm intelligence. Each type of algorithm has its own strengths and weaknesses, but all attempt to learn about, and exploit, consistent patterns in the data on which they operate. CI algorithms aim to solve problems more or less in the way that humans do but with the ability to deal with much larger and more complex datasets than could ever be manipulated by a human mind. CI algorithms often produce solutions that humans find nonintuitive and that are therefore potentially valuable, particularly in a field such as synthetic biology in which novel solutions are at a premium.

CI algorithms were developed specifically to deal with complex, data-rich problems in domains that are poorly understood. Most of these algorithms are highly stochastic and depend upon heuristics and therefore cannot be guaranteed to produce the 'best' solution to a given problem. They are valuable where a 'good enough' solution is acceptable or where multiple solutions can be analysed to provide insight into the underlying problem domain: the so-called *ensemble* approach. CI of all flavours has a lot to offer synthetic biology, but many potentially useful algorithms have not yet been explored in the context of synthetic biology. Undoubtedly, the most widely used CI algorithms to date are EAs.

3.1 Evolutionary algorithms

Evolutionary computation (EC) has been around since before computers were a consumer item (Box, 1957). EC attempts to mimic the elements of biological evolution to generate solutions to complex problems. Since a fitness function can be set up to reflect only the quality of a problem solution, rather than how the problem is solved, EC is ideally suited to problems in complex, poorly understood domains, where a good (not necessarily optimal) solution is essential but the precise nature of the

---

[8] http://www.inderscience.com/jhome.php?jcode=IJCIBSB
[9] http://www.ripublication.com/jcib.htm
[10] http://www.cibcb.org/
[11] http://www.cib-2012.org/
[12] http://prib.i3s.unice.fr/

3 Computational Intelligence

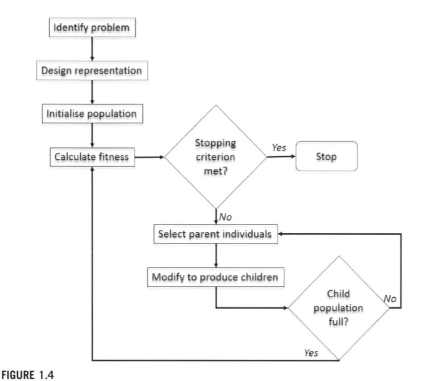

FIGURE 1.4

The simple genetic algorithm, one of the most widely used variants of EC.

solution is not crucial. There are many variants of EC (Goldberg, 1989; Mitchell, 1996; Fogel, 1999), including some specifically designed for bioinformatics (Fogel et al., 2008), but the basic principles are common to them all (Figure 1.4).

EAs are based, in a very simplistic manner, upon biological evolution. The first design decision to be made when developing an EA is to identify a means of representing a potential solution in a manner that can be read and manipulated by a computer. Consider a very simple, toy problem: the 'Hello World' of EAs.[13] Imagine that, for some reason, given a string of 0s and 1s, we wish to produce a string consisting solely of 1s. If the string is n characters long, our solution representation is simply a string of n digits (Figure 1.5).

A population of p potential solutions is generated by producing p strings, each of length n, with each position randomly assigned to 1 or 0. The first step in the EA is to assess the *fitness* of each string, using a *fitness function*: a function that takes a

---

[13]Traditionally, when a programmer learns a new programming language, the first programme written produces the output 'Hello World!'.

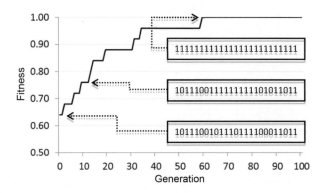

FIGURE 1.5

Course of evolution of an EA aimed to achieve a string of 1s. In this particular algorithm, the initial solution is generated at random, the length of the candidate solution is 25, and the mutation rate is 1 bit per generation. The fittest solution is always preserved, so fitness never declines. These design decisions do not hold for all EAs.

potential solution as input and produces a single number as output. For our problem, the fitness function can simply output the number of 1s in the string.

The next step is to select parent solutions to contribute to the next generation. Parents are chosen in a fitness-proportionate manner, so that any solution, no matter how poor, has a chance of contributing to the next generation, but fitter solutions are more likely to be selected. Once parents are selected, they are subjected to any combination of a range of evolutionary operators. Common evolutionary operations include *mutation*, in which bits of the solution are flipped at random, with a low probability (1–0 or 0–1), and crossover, where a breakpoint is selected and the first and second fragments of the two parents are swapped. Offspring generated in this manner are added to the next generation, and when the child population is full, the fitness of each member is assessed. This process iterates until a *stopping criterion* is met. The stopping criterion might be a specific number of generations, a particular fitness or a plateau in fitness that lasts for a set number of generations. In our example, the maximum fitness, a string of all 1s, with a fitness of 25, was achieved in just under 60 generations.

This EA is clearly a very simple one, and there are numerous variations on the basic theme, with different strengths and weaknesses. For some useful reviews, see Hallinan and Wiles (2002) and Zhou et al. (2011), and specifically, for network design Rodrigo, Carrera, and Elena (2010).

From the point of view of a synthetic biologist, EAs have a number of advantages:

- EAs facilitate genuine top-down design. With an EA, all that is needed is a set of reasonably well-characterised, computationally accessible parts, a method of combining them into a circuit and a fitness function. One readily implemented fitness function is a measure of the difference between a specific phenotype, or pattern of behaviour, and the behaviour of a model of the system. SVPs can be

combined into a simulateable model of a circuit, which can be run, and its output compared with the desired behaviour. With an EA, it is not necessary to know full details of all possible interactions between every possible pair of parts; dynamic behaviour should emerge from the simulation of the model.
- EAs are highly stochastic, and multiple runs of an algorithm will produce multiple solutions of equivalent fitness. Not all of these solutions will be intuitively obvious, or even biologically plausible, but the generation of multiple solutions can be a valuable approach to hypothesis generation.
- The use of an EA moves the intellectual input of the designer from the low-level assembly of parts to a high-level assessment of the biological implications of specific combinations of genetic elements and can therefore be a more productive use of the designer's time than a bottom-up design.

There are, of course, drawbacks to the EA approach. A naïve algorithm, which relies only upon the combination of already well-characterised parts, is restricted in its exploration of solution space. As discussed earlier, even for small numbers of parts, the set of potential circuits can be large, and solution space grows exponentially with the number of parts available. Under these circumstances, the number of potential solutions for any given phenotype is immense, but an EA can only explore that subset of the space that is encompassed by the parts available to it. Some of the more sophisticated algorithms can overcome these issues to some extent. For example, an algorithm can modify characteristics of the parts, such as parameters and interactions, within the model, starting from characterised parts. The resulting computationally evolved parts then provide templates for the development of new biological parts to be designed *de novo*, but this is a process that is currently not straightforward.

The applicability of EC to the design of genetic circuits is clear. As already discussed, our understanding of the details of possible interactions between biological parts (and their virtual computational counterparts) is woefully incomplete. However, synthetic biologists, of necessity, generally have a fairly clear idea of the behaviour they wish to produce. Assuming that the amount of knowledge built into virtual parts is adequate—a matter that is currently the subject of ongoing research—interesting behaviour can be allowed to emerge in the course of model simulation and be selected for or against as necessary.

The value of EC has previously been recognised by practitioners of metabolic engineering, a field closely allied to synthetic biology (Patil, Rocha, Forster, & Nielsen, 2005). Tasks to which these algorithms have been applied include pathway optimisation to maximise the production of a target biochemical (Mendes & Kell, 1998; Patil et al., 2005; Rocha et al., 2010) and the optimisation of parameters for allosteric regulation of enzymes (Gilman & Ross, 1995).

Metabolic engineers routinely transfer genes and pathways between organisms. Under these circumstances, it is often valuable to optimise the DNA sequence for expression in the new host. Non-native genes may be poorly expressed, even by otherwise reliable workhorse organisms (Angov, 2011). An important consideration is the refactoring of codon usage.

Codon usage is usually modified by mimicking the gene characteristics thought to be important for expression or by copying the codon usage of existing genes that are highly expressed in the new host. However, a more data-driven approach, based on CI, has been shown to be highly effective in increasing expression levels (Welch et al., 2009). Welch and colleagues used a Monte Carlo algorithm based on a codon frequency look-up table to generate sets of candidate sequences for two genes to be expressed in *E. coli*: the DNA polymerase of *Bacillus* phage Φ29 and a synthetic single-chain antibody fragment. A variety of different look-up tables and constraints were applied to create variant designs that differed in a number of parameters previously associated with expression effects in the literature. These factors included GC content, codon adaptation index (Sharp & Li, 1987), the presence of rare codons and the presence of specific RNA structures close to the site of initiation of transcription. A genetic algorithm incorporating mutation and crossover operators was used to select the codon subsets that were most predictive of expression levels. The authors found that synonymous substitutions can cause consistent, predictable expression differences, with up to 40-fold variation in expression. Interestingly, the preferred codons were not necessarily those used most frequently by the host organism. Further, variance in expression levels appears to involve features distributed across the CDS, rather than individual codon usages. These insights would not have emerged without the use of data-driven, hypothesis-free data mining.

Within synthetic biology, EAs have been applied to the design of a range of different types of genetic circuit (Table 1.1).

As Table 1.1 illustrates, most of the identified problems have been addressed repeatedly by different researchers, using different approaches. As a brief example, consider the design of a circuit that functions as an oscillator. Periodic oscillations in the levels of various proteins are a widespread phenomenon in most organisms, including bacteria, both within (Lenz & Søgaard-Andersen, 2011) and between (Mina, di Bernardo, Savery, & Tsaneva-Atanasova, 2013) cells, and there has been widespread interest in the design and implementation of synthetic oscillatory circuits (Hasty, Dolnik, Rottschäfer, & Collins, 2002; Stricker et al., 2008; Tigges, Marquez-Lago, Stelling, & Fussenegger, 2009; Danino, Mondragón-Palomino, Tsimring, & Hasty, 2010; Purcell, Savery, Grierson, & di Bernardo, 2010; Mondragón-Palomino, Danino, Selimkhanov, Tsimring, & Hasty, 2011). The first, and still the best-studied, synthetic oscillator is a manually designed circuit known as the *repressilator* (Elowitz & Leibler, 2000). This circuit consists of three genes, each of which represses one of the others, plus GFP as a reporter protein. The circuit is implemented on two plasmids and has been demonstrated to work *in vivo* more or less as designed.

The repressilator was designed manually. To design an oscillator using EC, the practitioner must first find a problem representation that is amenable to computational manipulation. A fitness function must then be devised to reflect the degree of oscillation of activity over time of one or several nodes. The algorithm can then be run iteratively until it converges, and the fittest solution accepted. Multiple runs of the algorithm will produce different, equally fit, solutions that can be compared for efficiency, cost and practicality of implementation and other factors.

Table 1.1 General Categories of Circuit Design to Which EAs Have Been Applied

Problem	Reference
Mathematical functions	Deckard and Sauro (2004)
	Miller, Job, and Vassilev (2000)
Network motifs	Paladugu et al. (2006)
	Kashtan and Alon (2005)
Logic gates[a]	Marchisio and Stelling (2011)
	Rodrigo and Jaramillo (2008)
	Ali, Almaini, and Kalganova (2004)
	Dari, Kia, Wang, Bulsara, and Ditto (2011)
	Coello Coello and Aguirre (2002)
Bistable switches	Francois and Hakim (2004)
Oscillators	Francois and Hakim (2004)
	Rodrigo, Carrera, and Jaramillo (2008)
	Rodrigo, Carrera, and Jaramillo (2011)
	Jin and Meng (2011)
Genotype–phenotype mapping	Guido et al. (2006)
	Welch et al. (2009)
Memory devices	Rodrigo and Jaramillo (2008)

This table provides an overview only; more detail is provided in the text.
[a]The literature on the evolutionary design of logic gates is extensive and beyond the scope of this review to cover in full. The references provided are amongst those most relevant to synthetic biology.

3.1.1 Modularity in evolutionary design

Synthetic genetic circuits are usually conceived of as existing in glorious isolation, performing clearly identifiable tasks. At present, however, we have neither the knowledge nor the technical ability to design entire genomes from scratch and understand all of the interactions that will occur between components. One way of overcoming the daunting complexity of genome-scale systems design is to break the system down into more or less self-contained modules. Biological systems are naturally modular (Hartwell, Hopfield, Leibler, & Murray, 1999), at a range of scales (Hallinan, 2004), and incorporating modularity from the very beginning of the synthetic design process means that designers can work at a higher, functional, level of abstraction than would otherwise be possible: considering, for example, 'an oscillator' instead of 'three promoters, an RBS, three CDS and two terminators, implemented on two plasmids'.

Although naturally occurring biological pathways, like synthetic circuits, are usually studied in isolation, there is evidence that a considerable amount of crosstalk occurs between pathways (Natarajan, Lin, Hsueh, Sternweis, & Ranganathan, 2006), a phenomenon that is clearly of value to the cell. However, in synthetic biology, crosstalk is generally regarded as undesirable and likely to lead to unintended phenotypic consequences. Synthetic circuits are generally designed to be *orthogonal*, as far as is possible, to existing chassis components (Benner, 2004; Benner & Sismour, 2005; Rackham & Chin, 2005; Channon, Bromley, & Woolfson, 2008; Wang, Kitney, Joly, & Buck, 2011).

One of the earliest applications of EAs to the design of genetic circuits was the evolution of oscillators (Francois & Hakim, 2004). These authors managed to evolve a variety of designs for each behaviour and found that posttranscriptional modifications were crucial to obtaining the functionality they sought. This observation highlights one of the weaknesses of any modelling approach to genetic circuit design: biological behaviours (posttranscriptional and posttranslational modifications, environmental effects, interactions with other pathways and so on) that are not included in the model cannot be used in the design. Had François and Hakim not made this option available to their algorithm, they would have produced different, and possibly less effective, designs.

The concept of using a small number of interacting genes to form a module that performs a specific function, and then combining modules to form larger, more complex circuits, is attractive and has clear parallels with the design of the silicon chips used in electronic circuit design. Statistical techniques for the identification of 'overrepresented' network motifs were developed by the systems biology community at the beginning of the twenty-first century (Milo et al., 2002) and have been the subject of intensive research ever since (for technical reviews of motif detection algorithms, see Ciriello and Guerra (2008) and Wong, Baur, Quader, and Huang (2012)). The underlying idea is that these motifs are overrepresented because they have been under positive selection pressure and therefore must have some specific biological function. In electrical circuits, motifs like these are used to control factors such as the amplitude of an electrical current; in biological circuits, they are thought to control the expression levels of one or more proteins.

Widely investigated motifs include, amongst many others, feed-forward loops and bi-fans in which two regulators cross regulate two target proteins (Figure 1.6).

The genes, or their products, in a network motif engage in biological interactions such as transcriptional control. The nature of an interaction—excitatory or inhibitory—affects the behaviour of the motif. In Figure 1.6, the designated 'output' protein of the feed-forward loop would be Z, while those of the bi-fan are Z and W. The feed-forward loop has been hypothesised to act as a delay element in transcriptional networks (Mangan, Zaslaver, & Alon, 2003). The bi-fan motif, in which two transcription factors each control two products, appears, despite its ubiquity, to have behaviour that is not so clearly definable (Ingram, Stumpf, & Stark, 2006).

Network motifs have been investigated both as a means for understanding existing systems and as a means of identifying potential building blocks for synthetic circuits. While the motif concept is appealing, the underlying assumption that these biological modules can be combined in arbitrary ways has not yet been tested *in vivo*. Consequently, there is some debate about whether genetic motifs actually represent units of selection and therefore whether they act *in vivo* as they are designed to do *in silico* (Artzy-Randrup, Fleishman, Ben-Tal, & Stone, 2004; Hallinan & Jackway, 2005; Ingram et al., 2006; Knabe, Nehaniv, & Schilstra, 2008). Despite these drawbacks,

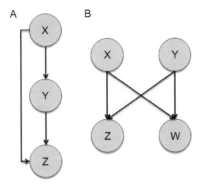

FIGURE 1.6

Examples of network motifs. (A) A feed-forward loop. (B) A bi-fan. Circles represent genes or gene products, while arrows represent regulatory interactions. (For colour version of this figure, the reader is referred to the online version of this chapter.)

small functional modules have been the subject of considerable research in synthetic biology (Paladugu et al., 2006; Hallinan & Wipat, 2007; Purnick & Weiss, 2009).

EAs have been applied to the development of network motifs with a variety of behaviours, using a variety of modelling approaches. Paladugu and colleagues, for example (Paladugu et al., 2006), used three different formalisms for modelling their networks: mass action kinetics (Kee, Coltrin, & Glarborg, 2005), Michaelis–Menten kinetics (Cornish-Bowden, 1995), and Hill kinetics (Hill, 1910), all modelled using ODEs. Starting from random initial networks of varying sizes, they evolved motifs to perform functions such as bistable switches, oscillators, maintenance of homeostasis and frequency filters (low-pass, high-pass, and band-pass). The fitness value of each network was calculated by selecting one node at random to be the output node and one to be the input node and then solving the equations representing the network using fourth-order Runge–Kutta integration (Butcher, 1987) with a fixed step size. Mutation was applied both to the network topology and to the values of the kinetic parameters. Motifs with a variety of topologies and simulated behaviours were identified, but were not implemented *in vivo*.

With the advent of the ubiquitous digital desktop computer, computation has become a widespread metaphor for the function of cells. The emphasis upon the role of cells as the fundamental building blocks of multicellular organisms, first proposed by Schleiden and Schwann in 1838, has shifted to a view of the cell as an information storage, processing and transmission machine, operating analogously to a computer. Referring to the *E. coli* chemotaxis network, Deckard and Sauro (2004) state this view explicitly: "Essentially the circuit acts as a simple but effective analogue computer".

Logic gates, the basis of electronic engineering, can be represented as a subset of network motifs. Logic gates are simple devices that produce a specific output in response to each possible combination of input signals and can be chained together to perform more complex calculations (Figure 1.7). Logic gates can be implemented

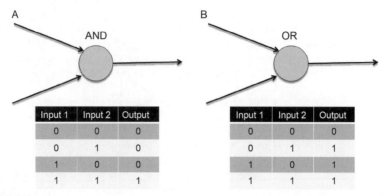

FIGURE 1.7

Examples of logic gates. (A) An AND gate produces an 'on' output (1) only when all of its inputs are 'on'. (B) An OR gate produces an 'on' output when any of its inputs are 'on'. The possible combinations of inputs, and the corresponding outputs, are represented in the form of truth tables below the gate diagram. (For colour version of this figure, the reader is referred to the online version of this chapter.)

in vivo using combinations of promoters and RBSs as inputs and the production of an RNA as output. Again, these circuits are very small, and the reasoning behind the interest in logic gate evolution is that it may be possible to combine such circuits *in vivo* as they are *in silico*.

EAs have been extensively applied to the optimisation of logic gates. Initially, practitioners of EC were simply fascinated by the possibilities of their field, and the value of evolving computational representations of logic gates appeared self-evident to these computer scientists, engineers and physicists. These early models were purely computational. With the advent of systems and then synthetic biology, modellers began to incorporate into their algorithms factors that would facilitate *in vivo* implementations of their designs (Rodrigo & Jaramillo, 2008).

For example, Marchisio and Stelling evolved very small circuits implementing logic gates (Marchisio & Stelling, 2011). Parameter optimization was carried out using a standard genetic algorithm. The issue of implementation practicality was addressed by including an additional fitness function to select between the numerous candidate circuit designs. This fitness function was a complexity score based upon the constraints of an *in vivo* implementation.

The view of cells as computational devices inspired some of the earliest work on the *in silico* evolution of genetic circuit designs. Circuits have been evolved to perform mathematical operations such as multiplication, square- and cube-root calculators and logarithmic functions (Deckard & Sauro, 2004), although these networks were not implemented *in vivo*. These authors developed an ODE for each node in the network and solved the system of ODEs using fourth-order Runge–Kutta (Butcher, 1987) techniques. Multiple solutions to some of the problems, such as the square-root solver, were found, in a relatively small number of generations.

Simple networks exhibiting bistable switch behaviour were first produced using computational evolution by Francois and Hakim (Francois & Hakim, 2004). These authors used a simple EA operating upon a population of candidate gene network models. As well as the promotion and repression of transcription, proteins in their model could interact to form complexes and could be posttranscriptionally modified. Posttranscriptional modification is known to be an important component of genetic regulatory systems. The details of how most posttranscriptional modification affects cellular function are still obscure for most systems, and hence, these details are rarely included in gene network models. Francois and Hakim's networks were subjected to a range of 'mutation operators' including the modification of kinetic constants and addition and deletion of genes and were scored according to how closely their behaviour matched the target behaviour. They found that bistable switches could be evolved in less than 100 generations, while oscillatory circuits could be achieved in a few hundred generations.

Francois and Hakim's results are particularly interesting because they illustrate the variety of network topologies that can produce bistable switch behaviour. Some of the bistable switch network designs were very similar to known biological switches such as the *lac* operon (Jacob & Monod, 1961), while others were novel. The networks also tended to be relatively robust to variations in parameter values and to noise. This robustness is a good indication that it is the network topology, rather than specific kinetic values, that is the most important factor in producing the observed behaviour. The relationship between network topology and dynamics is still not well understood, although there is considerable ongoing work in this area (Yang, Senthilkumar, Sun, & Kurths, 2011; Chalancon et al., 2012; Mejia-Guerra, Pomeranz, Morohashi, & Grotewold, 2012). Francois and Hakim's work is one demonstration of the way in which EC approaches can produce insights into these interactions.

In order to advance the field of synthetic biology, insights such as these must be implemented as useful biological systems. Computational and laboratory approaches to the design of living systems are essential complements to each other. Much of the EC work carried out to date in synthetic biology has not led to physical implementation, although some workers have taken this step.

Computational modelling was tightly integrated with laboratory experimentation in an elegant set of experiments carried out by Guido and colleagues (2006), aimed at quantifying the relationship between circuit topology and the dynamics of protein expression. These experiments involved using phage λ promoters inserted into the *lac* operon in *E. coli*. In order to examine the behaviour of the unregulated system, the researchers started by engineering the system in such a way that no regulation occurred. They used data on the transcriptional activity of this system to build a least-squares mathematical model of the system, parameterised using the best of 2000 random initial guesses for the model parameter values. The biological system was then modified to incorporate various combinations of gene activation and repression, and the model parameters were fitted to these new datasets. The original deterministic model was next modified to incorporate stochastic noise, and it was observed that fluctuations in the concentrations of transcription factors had very minor effects on the variability of expression levels. The model was used to generate testable predictions as to how changes in the system affected expression levels. The researchers concluded that their simple model had captured many, but not all, of the sources of variability in the biological system.

This work is interesting for several reasons. The use of tightly integrated, iterative modelling and experimental processes maximises the amount of information gained from both the computational and laboratory experiments and is a good model for future progress in synthetic biology. The work also demonstrates the potential of combining computational models of regulatory subsystems to predict the behaviour of larger, more complex networks. The goal of much systems biology research into computational 'virtual cells' has long been the development of sophisticated models of biological subsystems, which can be combined into models of the entire cell (Tomita, 2001; Joyce & Palsson, 2006; Karr et al., 2012). This approach is considerably more sophisticated than the search for network motifs, which has been the focus of most research in this area, and may have much to offer synthetic biology. This approach is likely to be even more productive for synthetic than for systems biology, since synthetic biology researchers are not restricted to the use of existing biological modules. However, the design of completely new components and models from scratch does, perforce, bring its own suite of challenges.

3.1.2 Simulated annealing

A widely used variant of the standard genetic algorithm is simulated annealing (SA), a generally applicable approach to the solution of combinatorial optimisation problems (van Laarhoven & Aarts, 1987). It is based on an analogy with the physical annealing of solids. Annealing is a process whereby a solid in a heat bath is heated to a temperature at which all the particles of the solid are randomly organised as a liquid. The temperature of the system is then gradually lowered, and the particles arrange themselves in the lowest energy state as an orderly lattice. At each temperature value T, the solid is allowed to reach thermal equilibrium, characterised by a probability of being in a state with energy E.

For a combinatorial optimisation problem such as kinetic parameter optimisation, the set of potential solutions assumes the role of the states of a solid, the cost function (or fitness function) replaces energy, and the control parameter, T, replaces temperature. The goal of the procedure is then to achieve a state of minimum cost (corresponding to maximum fitness) by moving between solutions with a probability that is dependent upon the 'temperature' of the system. At a high temperature, a higher cost solution is more likely to be accepted, while as the temperature decreases, the probability of accepting a higher cost move also decreases. The temperature is lowered during the running of the algorithm according to a predetermined schedule (van Laarhoven & Aarts, 1987).

The algorithm therefore traverses the search space as follows. Initially, the temperature is high and the cost function is desensitised and the current solution tends to

3 Computational Intelligence

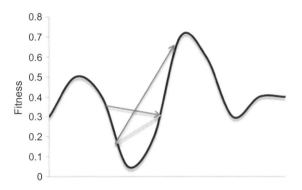

FIGURE 1.8

Simulated annealing. Trajectory of the algorithm on a hypothetical 1D fitness surface. The temperature of the system is indicated by the colour of the line at each step from high (red) to low (blue). (See color plate.)

'bounce around' the search space with little regard for the cost of each move. In this stage, the aim is to find the widest (which is assumed to be the deepest) 'valley' in the search space. As the temperature is decreased, the probability of moving to a point of higher cost becomes lower, and the algorithm performs a more restrained, local search. In the limit, as the temperature approaches 0 and time approaches infinity, the search converges to the global minimum (Figure 1.8).

It has been proven that, under specific temperature reduction conditions, SA is guaranteed to converge to the global minimum. Unfortunately, this temperature reduction is normally too slow for practical applications. With realistic temperature reduction schedules, one cannot assume that SA will always find the optimum solution for a given problem. Despite this, SA has been applied to many problem domains, with considerable success. There are two major problems with the algorithm: it tends to be very slow and a realistic temperature schedule must be established, usually by trial and error.

Several CAD tools for synthetic biology incorporate SA, including Genetdes (Rodrigo, Carrera, & Jaramillo, 2007) and SynBioSS (Weeding et al., 2010). SA has been applied to problems such as parameter optimisation for the repressilator network, allowing the system to be set to give oscillations of an arbitrary specified period (Tomshine & Kaznessis, 2006), and the design of synthetic ribosome binding sites (Salis, Mirsky, & Voigt, 2009).

3.1.3 The practice and promise of EAs for synthetic biology

There are two striking generalities evident in the synthetic biology evolutionary design literature. Firstly, the rapidity with which this approach to design was taken up: the term 'synthetic biology', coined in 1912 by the French chemist Stephane Leduc (De Lorenzo & Danchin, 2008), was first applied to the field as we know it in the early twentieth century (Endy, 2005),[14] and EAs were already being used in the field at that time (Deckard & Sauro, 2004).

The second striking feature of this subfield is the generally small size of the circuits produced. Although the ability to design large, complex circuits is often cited as a reason for the use of CAD and design automation (Barrett et al., 2006; Marchisio & Stelling, 2009; Esvelt & Wang, 2013), in practice, the circuits that have been published consist of only small numbers of genes. This limitation is particularly true for those circuits that have been implemented *in vivo*. The lack of large-scale designs may be due to the amount of computational resources needed to simulate genome-scale models; the relative slowness of the field in taking up the computational tools needed; to a desire on the part of the designers to fully understand the workings of the circuits they produce; to the practical problems of implementing even moderately sized circuits in living bacteria; or to a combination of factors.

There are a number of potentially important commonalities that emerge from this body of work. All of the projects use a simulateable model of a synthetic circuit to produce the data from which the fitness function is calculated. This approach is also widely used by the metabolic engineering and biotechnology communities (Lewis, Nagarajan, & Palsson, 2012; Branco dos Santos, de Vos, & Teusink, 2013) and has proven valuable in reducing the time and cost required to produce engineered organisms.

Ultimately, synthetic biologists are more interested in the performance of a genetic circuit than its inner workings, except insofar as these inner workings affect the practical implementation of a design *in vivo*. Those projects in which designs have actually been implemented in a bacterium highlight the gap between the complexity of a model and that of it equivalent DNA sequence. Most designs implemented have involved only small numbers of genes, but even small designs can require sophisticated cloning strategies. The recent increase in efficiency and fall in cost of DNA sequencing should address this problem to some extent, although long sequences can be expensive to synthesise and some sequence characteristics, such as the presence of low-complexity repeats, can make synthesis technically challenging.

Annotation is another important issue. In order to produce novel, emergent behaviour in a simulated, and eventually an implemented, circuit, the parts used to construct the model must somehow incorporate an adequate richness of information. One approach—the one that is currently most widely used—is for the designer to build in this knowledge, for example, by selecting the appropriate parts and constraining their interactions. This approach is slow, tedious and error-prone, and most importantly, it does not permit sharing of information

---

[14]Many practitioners claim to be have been carrying out research in the field for many years previous to 2000. For interesting discussion on the nature and history of synthetic biology see De Lorenzo and Danchin (2008) and Campos (2010).

between designs. The alternative, as discussed earlier, is to annotate individual parts with the required data, in a standard format. Again, this process is initially slow and painstaking, although there has been some research into the automated annotation of computational models (Lister, Pocock, Taschuk, & Wipat, 2009). Parts annotation, however, has the immense advantage that once a part is annotated, and stored in a shared repository, it can be used in any future design without further modification. The feasibility of this approach depends upon the existence of appropriate standards and databases and their adoption by the synthetic biology community.

Circuit designs generated by EAs also tend to be unnecessarily complex, and some authors have resorted to manual pruning of the designs (Deckard & Sauro, 2004; Paladugu et al., 2006) or the incorporation of a complexity penalty into the fitness function (Marchisio & Stelling, 2011). The positive aspect of this design exuberance, however, is the fact that multiple runs of an algorithm will produce multiple designs. Analysis of recurring features, such as network motifs, can provide insights into the identification of biological functionality such as possible functional modules and their interactions.

Overall, it is clear that EC is potentially valuable for the development of complex, novel, nonintuitive designs for synthetic genetic circuits. To date, this promise has not been fully realised; the circuits produced tend to be small and are selected because the researchers involved are interested in the theoretical aspects of behaviours such as switching, oscillations or logic gates, rather than the production of useful phenotypes.

3.2 Other CI techniques

There are several other classes of algorithm that are generally considered to be CI techniques. Nearly all of them have been applied to the analysis of microbial data at some time (for more details, see Hallinan (2012)). However, to date, none appear to have been applied to the design of genetic circuits, although some have been applied to investigations relevant to synthetic biology. They are described briefly here for completeness, because they may, in the future, prove to be useful to this rapidly developing discipline.

3.2.1 Neural networks

ANNs are an approach to machine learning and classification, based loosely upon the structure of biological brains (Haykin, 1994). ANNs are, indeed, very powerful classifiers; it has been mathematically proven that an ANN can learn any mathematical function to arbitrary precision, given enough training time and sufficient data (Hecht-Nielsen, 1989).[15]

---

[15]There is, however, no way of determining how much data or computational time is required to solve any given problem, or even whether there is sufficient data or time in the universe.

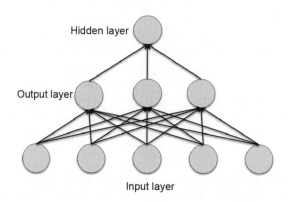

FIGURE 1.9

A simple fully connected feed-forward neural network with an input layer consisting of five nodes, one hidden layer of three nodes and an output layer of one node.

ANNs are built up of *nodes*, representing neurons, and *connections* between nodes, representing axons and dendrites carrying information. Connections in an ANN are weighted. The neurons are organised in layers, with an *input layer* representing one particular input data vector, an *output layer* providing a result of the classification and one or more *hidden layers*, so called because they are not connected to the outside world. The simplest and by far the most widely used form of ANN is the *perceptron*, a fully connected feed-forward network (Figure 1.9). Different numbers of nodes, and different patterns of connectivity can be used, but the vast majority of ANNs used for practical purposes are *perceptrons*, fully connected feed-forward networks (Figure 1.9; Sarle, 1994).

Each node in the hidden and output layers calculates the weighted sum of its inputs to determine the final input value. For a given node, j, the output value of each of the nodes, i, connected to node j is multiplied by the weight, $w_{i,j}$, on that connection. The values of all of the inputs are summed:

$$x = \sum_{i=1}^{n} (v_i w_{i,j}) \tag{1.1}$$

This value is then modified using a nonlinear *transfer*, or *squashing*, function to produce the node's output value (Figure 1.10).

In the case of hidden layer nodes, this output becomes an input into the next layer of nodes; in the case of output nodes, this value is the final output (Haykin, 2008). The transfer function is often a sigmoid, but any squashing-type nonlinear equation can be used. The transfer function ensures that the numbers being dealt with by the ANN remain within defined bounds, becoming neither too large nor too small to be manipulated.

ANNs must be trained using known, labelled data. At the beginning of the training process, the weights on the network are initialised to small, random values. The

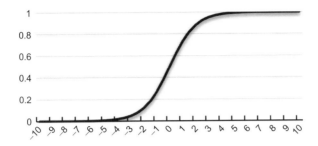

FIGURE 1.10

The sigmoid function. No matter how high or low the input value, the output is between 0.0 and 1.0.

objective of the training process is to modify the weights in such a way that they produce a consistent output for different classes of data: for example, if the aim is to classify proteins as 'secreted' or 'nonsecreted', the ANN could read in a vector of numbers representing characteristics of a protein—GC content, hydrophobicity, etc.—and output 0.0 for a secreted protein or 1.0 for a nonsecreted protein. This is the approach taken by tools such as SignalP, which uses a combination of ANNs and hidden Markov models to predict secreted proteins (Dyrløv Bendtsen, Nielsen, von Heijne, & Brunak, 2004).

The strength of ANNs lies in learning a mapping between inputs and outputs and being able to generalise this relationship to unseen data. The potential value of this power for learning the characteristics of existing, biological modules and extending the resulting knowledge to the design of new, modular synthetic systems has long been recognised (Hartwell et al., 1999; Purnick & Weiss, 2009).

ANNs have been applied to learning the mapping between the characteristics and the composition of bacterial community assemblages, both in terms of species abundance and of their distribution (Larsen, Field, & Gilbert, 2012). Using data on taxon abundances derived from measurements of 16S rRNA over 6 years, the researchers derived an environmental interaction network (EIN) in which nodes are microbial taxa and edges between them represent causal relationships. The EIN, in turn, was used to derive an ANN representation in the form of mathematical equations that best explain the data. This ANN was used to predict existing microbial taxa distributions, but these authors point out that this approach could also be used to design new ecosystems, by identifying the contribution of different factors to the make-up of the community. Synthetic ecosystems could be used, for example, to clean up contaminated sites or even to compensate for the effects of climate change.

Neural networks are powerful general-purpose learners of input/output mappings. They are widely used in a range of fields, including computational biology, but have not to date been widely used in the design of individual synthetic genetic circuits. When the synthetic biologist's focus moves to engineering populations, and even entire systems, these algorithms have been shown to be valuable. The design of

ANNs is something of a black art, and there are no rules as to the number and arrangement of neurons, connections, transfer functions and squashing functions needed to solve any given problem. In addition, ANNs generally require large amounts of labelled training data; the practitioners' rule of thumb is that the training dataset should contain at least five cases for each connection in the network. Despite these drawbacks, ANNs are powerful and flexible algorithms for classification and input/output mapping and deserve a place in the arsenal of any practitioner of CI.

3.2.2 Fuzzy systems

Fuzzy systems, as implied by their name, are a set of classification methods that attempt to divide the world into overlapping sets. Most classifiers will assign any given record to a single category: for example, the genetic and phenotypic characteristics of a particular bacterium may lead to it being designated as 'pathogenic' or 'nonpathogenic'. However, in the real world, such clear-cut classification is rarely the case: *Staphylococcus aureus* may be nonpathogenic if carried in the nasal passages but highly pathogenic if it manages to colonise the lungs or some other body site in which it does not normally occur. A human observer might classify *S. aureus* as, for example, 45% likely to be pathogenic.[16] In a fuzzy system, there may be any number of categories available and a single record will have a probability of membership of all of them. A fuzzy system boils down to a set of if–then rules that map inputs (characteristics of an individual) to outputs (class membership) (Kosko, 1994). As such, fuzzy systems are valuable in that they reflect the way in which humans think, but they can be complex to use because of their statistical characteristics.

In synthetic biology, fuzzy systems have, to date, been used to address the problem of parameter space searching for the identification of robust synthetic genetic circuits (Chen, Chang, & Lee, 2009; Chen & Wu, 2010). Fuzzy approaches have been widely applied to the field of ANNs, with promising results for learning in complex domains (Carpenter, Grossberg, Markuzon, Reynolds, & Rosen, 1992), and to other network analysis problems such as the identification of modules (Torra, 2002; Zhang, Wang, & Zhang, 2007; Horvath & Dong, 2008). In many domains, the assumption that a node must necessarily belong to only one module does not hold. A fuzzy systems approach to the design of network modules has the potential to address some of the issues of the interplay between modularity and crosstalk discussed earlier. To date, however, this approach does not appear to have been applied in circuit design.

3.2.3 Swarm intelligence

Swarm intelligence is based upon the principle of *self-organisation*, as observed in the social insects. Also known as ant-colony optimisation, this set of algorithms are predicated upon the observation that large numbers of simple agents, interacting with

---

[16] A number plucked completely out of thin air, for this example.

each other and their environment under the same, simple rules, can produce behaviour that is apparently purposeful and that can result in optimal solutions to a wide range of problems (Bonabeau, Dorigo, & Theraulaz, 1999). A variant of swarm intelligence is *particle swarm optimisation*, in which individual particles are potential solutions to a problem, in much the same way as individuals in an evolutionary algorithm. Particles move through multidimensional solution space and have both a position in that space and a velocity, determined by a set of rules (Kari & Rozenberg, 2008). Particle swarm systems have been used to model networks of interacting oscillators (Olfati-Saber, 2006) and for the design and control of complex biological systems (Jacob, 2008; Doursat, Sayama, & Michel, 2012). These applications, while interesting and of potential relevance to synthetic biology, are currently highly theoretical.

4 DISCUSSION

CI comprises a powerful suite of tools for exploring design space and mapping between inputs and outputs. The algorithms are designed to work in noisy, poorly understood domains, in which a 'good enough' solution is acceptable. CI has been applied successfully to domains as disparate as engineering, economics sociology and bioinformatics. It is somewhat surprising, then, that although EAs have been applied to the design of small genetic circuits ever since synthetic biology was acknowledged as a field in its own right, other CI approaches have been largely neglected. There are both advantages and disadvantages to the adoption of CI in synthetic biology (Table 1.2).

One of the weaknesses of any modelling-based approach to genetic circuit design is that biological phenomena that are not included in the model cannot be used in the

Table 1.2 A Summary of the Advantages and Disadvantages of the CI Approach to Synthetic Circuit Design

Pros	Cons
Can operate on poorly understood domains where 'good enough' solutions acceptable	Often produce a 'black box' input/output mapping
Many can work with incomplete data	May need large amounts of training data
Can produce nonintuitive designs	Designs may need to be optimised for *in vivo* implementation
Sometimes multiple solutions—fodder for further analysis and understanding of novel system	Search space can be very large, so long running times, incomplete coverage of solution space, need strategies for reduction
Inherently stochastic, hence good models of biological systems	

design. Factors such as posttranscriptional and posttranslational modifications, environmental effects and crosstalk with other pathways can have significant effects upon the rates of production of proteins and metabolites. The usual transcription-heavy approach to modelling incorporates the assumption that transcriptional control is the primary means of regulating protein levels in the cell, with 'strong' promoters leading to the production of more protein than 'weak' promoters (Alper, 2005). It is becoming clear that this simple relationship does not necessarily hold; indirect and suboptimal control of gene expression is widespread in bacteria (Price et al., 2013), and the relation between the amount of mRNA produced in an individual cell and the amount of its associated protein is not necessarily linear (Gygi, Rochon, Franza, & Aebersold, 1999). Indeed, mRNA abundance alone has been found to explain only 20–28% of the total variation of protein abundance in *Desulfovibrio vulgaris* (Nie, Wu, & Zhang, 2006). Although some workers are incorporating regulatory mechanisms such as RNA interactions, complex formation and epigenetic modifications into their designs, there is currently no clear consensus on the value of incorporating the various biological processes known to affect protein levels into the synthetic biology design task.

One way of addressing this fundamental lack of knowledge is to supplement *in silico* evolution with *in vivo*-directed evolution. A model that works as desired computationally can be implemented as DNA, inserted into a host bacterium, and then modified using directed evolution (Dougherty & Arnold, 2009; Brustad & Arnold, 2011; Cobb, Si, & Zhao, 2012). Since the sequence corresponding to the original design is known, a comparative analysis of the computationally and experimentally evolved sequences should provide some insight into the factors lacking in the original model, which can then be modified to reflect this new knowledge (Hallinan et al., 2012).

The use of CI techniques should also encourage designers to operate at a more global, whole-cell level than is currently the norm, since full information about a system is not required for some algorithms. For example, as discussed earlier, crosstalk between pathways is generally regarded as undesirable in a synthetic biology context. However, there is some evidence that, at least in mammalian cells, interpathway crosstalk occurs via a relatively small number of mechanisms(Natarajan et al., 2006), raising the intriguing, but as yet unpursued, possibility of synthetic biologists eventually being able to harness this aspect of biology to perform more complex, globally aware information processing.

In addition to engineering intracellular genetic circuits, EC has been used to design systems incorporating communication between individual bacteria. Such systems can be both more complex and more robust than individual-only systems, permitting the inclusion of multiple types of cells, behaving either in the same or in different ways. In order to achieve such a virtual multicellular system, most effort has been focussed upon engineering the quorum sensing systems of bacteria using the conventional, manual approach (Kobayashi, 2004; You, Cox, Weiss, & Arnold, 2004; Basu, Gerchman, Collins, Arnold, & Weiss, 2005). However, optimising and controlling the behaviour of a population of bacteria with differing phenotypes is a nontrivial task and one that cannot be performed without the aid of computational modelling and design (Brenner, You, & Arnold, 2008). CI approaches, particularly EC and swarm algorithms, offer a promising approach to this complex optimisation problem.

CI algorithms can be a valuable addition to the synthetic biologist's toolbox, but the algorithms must be chosen carefully to fit the characteristics of the problem. As with any algorithm, it is important to apply human, as well as CI to the circuit design task; computational algorithms can 'intelligent' in a very dumb way, and human sanity checking and occasional redirection will be an essential part of the synthetic biology design process for the foreseeable future.

References

Agapakis, C. M., Boyle, P. M., & Silver, P. A. (2012). Natural strategies for the spatial optimization of metabolism in synthetic biology. *Nature Chemical Biology*, *8*(6), 527–535.

Ali, B., Almaini, A., & Kalganova, T. (2004). Evolutionary algorithms and their use in the design of sequential logic circuits. *Genetic Programming and Evolvable Machines*, *5*(1), 11–29.

Alper, H. (2005). Tuning genetic control through promoter engineering. *Proceedings of the National Academy of Sciences*, *102*(36), 12678–12683.

Andrianantoandro, E., Basu, S., Karig, D. K., & Weiss, R. (2006). Synthetic biology: New engineering rules for an emerging discipline. *Molecular Systems Biology*, *2*: 2006.0028.

Angov, E. (2011). Codon usage: Nature's roadmap to expression and folding of proteins. *Biotechnology Journal*, *6*(6), 650–659.

Artzy-Randrup, Y., Fleishman, S. J., Ben-Tal, N., & Stone, L. (2004). Comment on "Network motifs: Simple building blocks of complex networks" and "Superfamilies of evolved and designed networks". *Science*, *305*(5687), 1107.

Barbe, V., Cruveiller, S., Kunst, F., Lenoble, P., Meurice, G., Sekowska, A., et al. (2009). From a consortium sequence to a unified sequence: The *Bacillus subtilis* 168 reference genome a decade later. *Microbiology*, *155*(6), 1758–1775.

Barrett, C. L., Kim, T. Y., Kim, H. U., Palsson, B.Ø., & Lee, S. Y. (2006). Systems biology as a foundation for genome-scale synthetic biology. *Current Opinion in Biotechnology*, *17*(5), 488–492.

Basu, S., Gerchman, Y., Collins, C. H., Arnold, F. H., & Weiss, R. (2005). A synthetic multicellular system for programmed pattern formation. *Nature*, *434*(7037), 1130–1134.

Benner, S. A. (2004). Understanding nucleic acids using synthetic chemistry. *Accounts of Chemical Research*, *37*(10), 784–797.

Benner, S. A., & Sismour, A. M. (2005). Synthetic biology. *Nature Reviews Genetics*, *6*(7), 533–543.

Blount, B. A., Weenink, T., Vasylechko, S., & Ellis, T. (2012). Rational diversification of a promoter providing fine-tuned expression and orthogonal regulation for synthetic biology. *PLoS One*, *7*(3), e33279.

Bonabeau, E., Dorigo, M., & Theraulaz, G. (1999). *Swarm intelligence: From natural to artificial systems*. New York: Oxford University Press.

Bongard, J. (2002). Evolving modular genetic regulatory networks. In *2002 IEEE conference on evolutionary computation*, San Diego, CA: IEEE Press.

Box, G. E. P. (1957). Evolutionary operation: A method for increasing industrial productivity. *Applied Statistics*, *6*(2), 81–101.

Branco dos Santos, F., de Vos, W. M., & Teusink, B. (2013). Towards metagenome-scale models for industrial applications—The case of Lactic Acid Bacteria. *Current Opinion in Biotechnology*, *24*(2), 200–206.

Brenner, K., You, L., & Arnold, F. H. (2008). Engineering microbial consortia: A new frontier in synthetic biology. *Trends in Biotechnology*, *26*(9), 483–489.

Brustad, E. M., & Arnold, F. H. (2011). Optimizing non-natural protein function with directed evolution. *Current Opinion in Chemical Biology*, *15*(2), 201–210.

Butcher, J. C. (1987). *The numerical analysis of ordinary differential equations: Runge–Kutta and general linear methods*. Chichester: Wiley-Interscience.

Campos, L. (2010). That was the synthetic biology that was. In *Synthetic biology* (pp. 5–21). Netherlands: Springer.

Canton, B., Labno, A., & Endy, D. (2008). Refinement and standardization of synthetic biological parts and devices. *Nature Biotechnology*, *26*(7), 787–793.

Carpenter, G. A., Grossberg, S., Markuzon, N., Reynolds, J. H., & Rosen, D. B. (1992). Fuzzy ARTMAP: A neural network architecture for incremental supervised learning of analog multidimensional maps. *IEEE Transactions on Neural Networks*, *3*(5), 698–713.

Carr, P. A., Wang, H. H., Sterling, B., Isaacs, F. J., Lajoie, M. J., Xu, G., et al. (2012). Enhanced multiplex genome engineering through co-operative oligonucleotide co-selection. *Nucleic Acids Research*, *40*(17), e132.

Carrera, J., Rodrigo, G., & Jaramillo, A. (2009). Towards the automated engineering of a synthetic genome. *Molecular BioSystems*, *5*(7), 733.

Chalancon, G., Ravarani, C. N., Balaji, S., Martinez-Arias, A., Aravind, L., Jothi, R., et al. (2012). Interplay between gene expression noise and regulatory network architecture. *Trends in Genetics*, *28*(5), 221–232.

Chandrakasan, A. P., Bowhill, W. J., & Fox, F. (2000). *Design of high-performance microprocessor circuits* (1st ed.). New York: Wiley-IEEE Press.

Chandran, D., Bergmann, F. T., & Sauro, H. M. (2009). TinkerCell: Modular CAD tool for synthetic biology. *Journal of Biological Engineering*, *3*(1), 19.

Channon, K., Bromley, E. H. C., & Woolfson, D. N. (2008). Synthetic biology through biomolecular design and engineering. *Current Opinion in Structural Biology*, *18*(4), 491–498.

Chen, D., & Arkin, A. P. (2012). Sequestration-based bistability enables tuning of the switching boundaries and design of a latch. *Molecular Systems Biology*, *8*, 620.

Chen, B.-S., Chang, C.-H., & Lee, H.-C. (2009). Robust synthetic biology design: Stochastic game theory approach. *Bioinformatics*, *25*(14), 1822–1830.

Chen, J., Densmore, D., Ham, T. S., Keasling, J. D., & Hillson, N. J. (2012). DeviceEditor visual biological CAD canvas. *Journal of Biological Engineering*, *6*(1), 1.

Chen, B.-S., & Wu, C.-H. (2010). Robust optimal reference-tracking design method for stochastic synthetic biology systems: T–S fuzzy approach. *IEEE Transactions on Fuzzy Systems*, *18*(6), 1144–1159.

Ciriello, G., & Guerra, C. (2008). A review on models and algorithms for motif discovery in protein–protein interaction networks. *Briefings in Functional Genomics & Proteomics*, *7*(2), 147–156.

Cobb, R. E., Si, T., & Zhao, H. (2012). Directed evolution: An evolving and enabling synthetic biology tool. *Current Opinion in Chemical Biology*, *16*(3–4), 285–291.

Coello Coello, C. A., & Aguirre, A. H. (2002). Design of combinational logic circuits through an evolutionary multiobjective optimization approach. *AI EDAM*, *16*(01), 39–53.

Cooling, M. T., Hunter, P., & Crampin, E. J. (2008). Modelling biological modularity with CellML. *IET Systems Biology*, *2*(2), 73–79.

Cooling, M. T., Rouilly, V., Misirli, G., Lawson, J., Yu, T., Hallinan, J., et al. (2010). Standard virtual biological parts: A repository of modular modeling components for synthetic biology. *Bioinformatics*, *26*, 925–931.

Cornish-Bowden, A. (1995). *Fundamentals of enzyme kinetics*. London: Portland Press.

Craddock, T., Harwood, C. R., Hallinan, J., & Wipat, A. (2008). e-Science: Relieving bottlenecks in large-scale genomic analyses. *Nature Reviews Microbiology*, *6*, 948–954.

Curbera, F., Duftler, M., Khalaf, R., Nagy, W., Mukhi, N., & Weerawarana, S. (2002). Unraveling the web services web. *IEEE Internet Computing*, *6*, 86–93.

Czar, M. J., Anderson, J. C., Bader, J. S., & Peccoud, J. (2009). Gene synthesis demystified. *Trends in Biotechnology*, *27*(2), 63–72.

Danino, T., Mondragón-Palomino, O., Tsimring, L., & Hasty, J. (2010). A synchronized quorum of genetic clocks. *Nature*, *463*(7279), 326–330.

Dari, A., Kia, B., Wang, X., Bulsara, A. R., & Ditto, W. (2011). Noise-aided computation within a synthetic gene network through morphable and robust logic gates. *Physical Review E*, *83*(4), 041909.

De Lorenzo, V., & Danchin, A. (2008). Synthetic biology: Discovering new worlds and new words. *EMBO Reports*, *9*(9), 822–827.

De Mey, M., Maertens, J., Lequeux, G. J., Soetaert, W. K., & Vandamme, E. J. (2007). Construction and model-based analysis of a promoter library for *E. coli*: An indispensable tool for metabolic engineering. *BMC Biotechnology*, *7*, 34.

Deckard, A., & Sauro, H. M. (2004). Preliminary studies on the *in silico* evolution of biochemical networks. *ChemBioChem*, *5*(10), 1423–1431.

Dougherty, M. J., & Arnold, F. H. (2009). Directed evolution: New parts and optimized function. *Current Opinion in Biotechnology*, *20*(4), 486–491.

Doursat, R., Sayama, H., & Michel, O. (2012). Morphogenetic engineering: Reconciling self-organization and architecture. In *Morphogenetic engineering* (pp. 1–24). Heidelberg: Springer.

Dyrløv Bendtsen, J., Nielsen, H., von Heijne, G., & Brunak, S. (2004). Improved prediction of signal peptides: SignalP 3.0. *Journal of Molecular Biology*, *340*(4), 783–795.

Edwards, R., & Glass, L. (2000). Combinatorial explosion in model gene networks. *Chaos*, *10*, 691–704.

Eiben, A. E., & Schippers, C. A. (1998). On evolutionary exploration and exploitation. *Fundamenta Informatica*, *35*, 1–15.

Eilbeck, K., Lewis, S., Mungall, C. J., Yandell, M., Stein, L., Durbin, R., et al. (2005). The Sequence Ontology: A tool for the unification of genome annotation. *Genome Biology*, *6*, R44.

Elowitz, M. B., & Leibler, S. (2000). A synthetic oscillatory network of transcriptional regulators. *Nature*, *403*, 335–338.

Endy, D. (2005). Foundations for engineering biology. *Nature*, *438*(7067), 449–453.

Engelbrecht, A. P. (2007). *Computational intelligence: An introduction*. Chichester: John Wiley.

Esvelt, K. M., & Wang, H. H. (2013). Genome-scale engineering for systems and synthetic biology. *Molecular Systems Biology*, *9*(1), 641.

Fernández-Suárez, X. M., & Galperin, M. Y. (2013). The 2013 Nucleic Acids Research Database Issue and the online Molecular Biology Database Collection. *Nucleic Acids Research*, *41*(D1), D1–D7.

Fogel, L. J. (1999). *Intelligence through simulated evolution: Four decades of evolutionary programming*. New York: Wiley.

Fogel, D., & Corne, D. W. (2003). *Evolutionary computation in bioinformatics*. Boston: Morgan Kauffman.

Fogel, G. B., Corne, D. W., & Pan, Y. (Eds.), (2008). *Computational intelligence in bioinformatics*. San Diego, CA: Wiley-IEEE Press.

Forster, A. C., & Church, G. M. (2006). Toward synthesis of a minimal cell. *Molecular Systems Biology*, *2*, 1–10.

Francois, P., & Hakim, V. (2004). Design of genetic networks with specified functions by evolution *in silico*. *Proceedings of the National Academy of Sciences of the United States of America*, *101*(2), 580–585.

Gibson, D. G., Glass, J. I., Lartigue, C., Noskov, V. N., Chuang, R.-Y., Algire, M. A., et al. (2010). Creation of a bacterial cell controlled by a chemically synthesized genome. *Science*, *329*(5987), 52–56.

Gilman, A., & Ross, J. (1995). Genetic-algorithm selection of a regulatory structure that directs flux in a simple metabolic model. *Biophysical Journal*, *69*(4), 1321–1333.

Goldberg, D. (1989). *Genetic algorithms in search, optimisation and machine learning*. Boston: Addison-Wesley.

Guido, N. J., Wang, X., Adalsteinsson, D., McMillen, D., Hasty, J., Cantor, C. R., et al. (2006). A bottom-up approach to gene regulation. *Nature*, *439*(7078), 856–860.

Gustafsson, C., Govindarajan, S., & Minshull, J. (2004). Codon bias and heterologous protein expression. *Trends in Biotechnology*, *22*(7), 346–353.

Gygi, S. P., Rochon, Y., Franza, B. R., & Aebersold, R. (1999). Correlation between protein and mRNA abundance in yeast. *Molecular and Cellular Biology*, *19*(3), 1720–1730.

Hallinan, J. (2004). Gene duplication and hierarchical modularity in intracellular interaction networks. *Bio Systems*, *74*(1–3), 51–62.

Hallinan, J. (2012). Data mining for microbiologists. *Methods in Microbiology*, *39*, 27–79 C. Harwood.

Hallinan, J., & Jackway, P. (2005). Network motifs, feedback loops and the dynamics of genetic regulatory networks. In *2005 IEEE symposium on computational intelligence in bioinformatics and computational biology*, San Diego, CA: IEEE Press.

Hallinan, J., Park, S., & Wipat, A. (2012). Bridging the gap between design and reality: A dual evolutionary strategy. BioInformatics: International Conference on Bioinformatics Models, Methods and Algorithms. Algarve, Portugal, ScitePress: 263–268.

Hallinan, J., & Wiles, J. (2002). Evolutionary algorithms. In L. Nadel (Ed.), *The encyclopedia of cognitive sciences*. New York: Palgrave Macmillan.

Hallinan, J., & Wipat, A. (2007). Motifs and modules in a fractured yeast functional network. In *2007 IEEE symposium on computational intelligence in bioinformatics and computational biology*, Honolulu, Hawaii: IEEE Press.

Hartner, F. S., Ruth, C., Langenegger, D., Johnson, S. N., Hyka, P., Lin-Cereghino, G. P., et al. (2008). Promoter library designed for fine-tuned gene expression in *Pichia pastoris*. *Nucleic Acids Research*, *36*(12), e76.

Hartwell, L. H., Hopfield, J. J., Leibler, S., & Murray, A. W. (1999). From molecular to modular cell biology. *Nature*, *402*, C47–C52.

Hasty, J., Dolnik, M., Rottschäfer, V., & Collins, J. J. (2002). Synthetic gene network for entraining and amplifying cellular oscillations. *Physical Review Letters*, *88*(14), 148101.

Haykin, S. O. (1994). *Neural networks: A comprehensive foundation.* Prentice Hall, NJ: Upper Saddle River.

Haykin, S. O. (2008). *Neural networks and learning machines: International version: A comprehensive foundation.* Upper Saddle River, NJ: Pearson.

Hecht-Nielsen, R. (1989). Theory of the backpropagation neural network. In *International joint conference on neural networks (IJCNN).* San Diego CA.

Heinemann, M., & Panke, S. (2006). Synthetic biology—Putting engineering into biology. *Bioinformatics*, 22(22), 2790–2799.

Hill, A. V. (1910). The possible effects of the aggregation of the molecules of haemoglobin on its dissociation curves. *Journal of Physiology*, 40(Suppl.), iv–vii.

Horvath, S., & Dong, J. (2008). Geometric interpretation of gene coexpression network analysis. *PLoS Computational Biology*, 4(8), e1000117.

Hucka, M., Finney, A., Sauro, H. M., Bolouri, H., Doyle, J. C., Kitano, H., et al. (2003). The systems biology markup language (SBML): A medium for representation and exchange of biochemical network models. *Bioinformatics*, 19, 524–531.

Ingram, P. J., Stumpf, M. P. H., & Stark, J. (2006). Network motifs: Structure does not determine function. *BMC Genomics*, 7, 108.

Jacob, C. (2008). Dancing with swarms: Utilizing swarm intelligence to build, investigate, and control complex systems. In *Design by evolution* (pp. 69–94). Berlin and Heidelberg: Springer.

Jacob, F., & Monod, J. (1961). Genetic regulatory mechanisms in the synthesis of proteins. *Journal of Molecular Biology*, 3, 318–356.

Jin, Y., & Meng, Y. (2011). Emergence of robust regulatory motifs from *in silico* evolution of sustained oscillation. *Bio Systems*, 103(1), 38–44.

Joyce, A. R., & Palsson, B.Ø. (2006). The model organism as a system: Integrating 'omics' data sets. *Nature Reviews Molecular Cell Biology*, 7(3), 198–210.

Kari, L., & Rozenberg, G. (2008). The many facets of natural computing. *Communications of the ACM*, 51(10), 72–83.

Karr, J. R., Sanghvi, J. C., Macklin, D. N., Gutschow, M. V., Jacobs, J. M., Bolival, B. Jr., et al. (2012). A whole-cell computational model predicts phenotype from genotype. *Cell*, 150(2), 389–401.

Kashtan, N., & Alon, U. (2005). Spontaneous evolution of modularity and network motifs. *Proceedings of the National Academy of Sciences of the United States of America*, 102(39), 13773–13778.

Kee, R. J., Coltrin, M. E., & Glarborg, P. (2005). Mass-action kinetics. In *Chemically reacting flow* (pp. 371–400). New York: Wiley.

Kelemen, A., Abraham, A., & Chen, Y. (Eds.), (2008). *Computational intelligence in bioinformatics. Studies in computational intelligence.* New York: Springer.

Klein, C., Kaletta, C., Schnell, N., & Entian, K. D. (1992). Analysis of genes involved in biosynthesis of the lantibiotic subtilin. *Applied and Environmental Microbiology*, 58(1), 132–142.

Knabe, J., Nehaniv, C., & Schilstra, M. (2008). Do motifs reflect evolved function?—No convergent evolution of genetic regulatory network subgraph topologies. *Bio Systems*, 94(1–2), 68–74.

Kobayashi, H. (2004). Programmable cells: Interfacing natural and engineered gene networks. *Proceedings of the National Academy of Sciences*, 101(22), 8414–8419.

Kosko, B. (1994). Fuzzy systems as universal approximators. *IEEE Transactions on Computers*, *43*(11), 1329–1333.

Larsen, P. E., Field, D., & Gilbert, J. A. (2012). Predicting bacterial community assemblages using an artificial neural network approach. *Nature Methods*, *9*(6), 621–625.

Lenz, P., & Søgaard-Andersen, L. (2011). Temporal and spatial oscillations in bacteria. *Nature Reviews Microbiology*, *9*(8), 565–577.

Lewis, N. E., Nagarajan, H., & Palsson, B. O. (2012). Constraining the metabolic genotype–phenotype relationship using a phylogeny of *in silico* methods. *Nature Reviews Microbiology*, *10*(4), 291–305.

Lin, L., & Gen, M. (2009). Auto-tuning strategy for evolutionary algorithms: Balancing between exploration and exploitation. *Soft Computing*, *13*(2), 157–168.

Lister, A. L., Pocock, M., Taschuk, M., & Wipat, A. (2009). Saint: A lightweight integration environment for model annotation. *Bioinformatics*, *25*(22), 3026–3027.

Mangan, S., Zaslaver, A., & Alon, U. (2003). The coherent feedforward loop serves as a sign-sensitive delay element in transcriptional networks. *Journal of Molecular Biology*, *334*, 197–204.

Marchisio, M. A., & Stelling, J. (2009). Computational design tools for synthetic biology. *Current Opinion in Biotechnology*, *20*(4), 479–485.

Marchisio, M. A., & Stelling, J. (2011). Automatic design of digital synthetic gene circuits. *PLoS Computational Biology*, *7*(2), e1001083.

Medema, M. H., van Raaphorst, R., Takano, E., & Breitling, R. (2012). Computational tools for the synthetic design of biochemical pathways. *Nature Reviews Microbiology*, *334*, 1716–1719.

Mejia-Guerra, M. K., Pomeranz, M., Morohashi, K., & Grotewold, E. (2012). From plant gene regulatory grids to network dynamics. *Biochimica et Biophysica Acta (BBA)-Gene Regulatory Mechanisms*, *1819*(5), 454–465.

Mendes, P., & Kell, D. (1998). Non-linear optimization of biochemical pathways: Applications to metabolic engineering and parameter estimation. *Bioinformatics*, *14*(10), 869–883.

Miller, J. F., Job, D., & Vassilev, V. K. (2000). Principles in the evolutionary design of digital circuits—Part I. *Genetic Programming and Evolvable Machines*, *1*(1–2), 7–35.

Milo, R., Shen-Orr, S., Itzkovitz, S., Kashtan, N., Chklovskii, D., & Alon, U. (2002). Network motifs: Simple building blocks of complex networks. *Science*, *298*(5594), 824–827.

Mina, P., di Bernardo, M., Savery, N. J., & Tsaneva-Atanasova, K. (2013). Modelling emergence of oscillations in communicating bacteria: A structured approach from one to many cells. *Journal of the Royal Society, Interface*, *10*(78), 20120612.

Misirli, G., Hallinan, J. S., Yu, T., Lawson, J. R., Wimalaratne, S. M., Cooling, M. T., et al. (2011). Model annotation for synthetic biology: Automating model to nucleotide sequence conversion. *Bioinformatics*, *27*(7), 973–979.

Mitchell, M. (1996). *An introduction to genetic algorithms*. Cambridge MA: MIT Press.

Mitra, S. (2005). Computational intelligence in bioinformatics. In J. Peters & A. Skowron (Eds.), *LNCS:* Vol. 3400. *Transactions on rough sets III* (pp. 134–152). Berlin Heidelberg: Springer.

Mondragón-Palomino, O., Danino, T., Selimkhanov, J., Tsimring, L., & Hasty, J. (2011). Entrainment of a population of synthetic genetic oscillators. *Science Signaling*, *333*(6047), 1315.

Natarajan, M., Lin, K.-M., Hsueh, R. C., Sternweis, P. C., & Ranganathan, R. (2006). A global analysis of cross-talk in a mammalian cellular signalling network. *Nature Cell Biology, 8*(6), 571–580.

Nie, L., Wu, G., & Zhang, W. (2006). Correlation between mRNA and protein abundance in *Desulfovibrio vulgaris*: A multiple regression to identify sources of variations. *Biochemical and Biophysical Research Communications, 339*(2), 603–610.

Noireaux, V., Maeda, Y. T., & Libchaber, A. (2011). Development of an artificial cell, from self-organization to computation and self-reproduction. *Proceedings of the National Academy of Sciences, 108*(9), 3473–3480.

O'Malley, M., Powell, A., Davies, J. F., & Calvert, J. (2008). Knowledge-making distinctions in synthetic biology. *BioEssays, 30*(1), 57–65.

Olfati-Saber, R. (2006). Swarms on sphere: A programmable swarm with synchronous behaviors like oscillator networks. In *45th IEEE conference on decision and control*.

Paladugu, S. R., Chickarmane, V., Deckard, A., Frumkin, J. P., McCormack, M., & Sauro, H. M. (2006). *In silico* evolution of functional modules in biochemical networks. *IEEE Proceedings. Systems Biology, 153*(4), 223–235.

Patil, K., Rocha, I., Forster, J., & Nielsen, J. (2005). Evolutionary programming as a platform for *in silico* metabolic engineering. *BMC Bioinformatics, 6*(1), 308.

Price, M. N., Deutschbauer, A. M., Skreker, J. M., Wetmore, K. M., Ruths, T., Mar, J. S., et al. (2013). Indirect and suboptimal control of gene expression is widespread in bacteria. *Molecular Systems Biology, 9*.

Purcell, O., Savery, N. J., Grierson, C. S., & di Bernardo, M. (2010). A comparative analysis of synthetic genetic oscillators. *Journal of the Royal Society, Interface, 7*(52), 1503–1524.

Purnick, P. E. M., & Weiss, R. (2009). The second wave of synthetic biology: From modules to systems. *Nature Reviews Molecular Cell Biology, 10*, 410–422.

Rackham, O., & Chin, J. W. (2005). A network of orthogonal ribosome-mRNA pairs. *Nature Chemical Biology, 1*(3), 159–166.

Rocha, I., Maia, P., Evangelista, P., Vilaça, P., Soares, S., Pinto, J., et al. (2010). OptFlux: An open-source software platform for *in silico* metabolic engineering. *BMC Systems Biology, 4*(1), 45.

Rodrigo, G., Carrera, J., & Elena, S. F. (2010). Network design meets *in silico* evolutionary biology. *Biochimie, 92*(7), 746–752.

Rodrigo, G., Carrera, J., & Jaramillo, A. (2007). Genetdes: Automatic design of transcriptional networks. *Bioinformatics, 23*(14), 1857–1858.

Rodrigo, G., Carrera, J., & Jaramillo, A. (2008). Computational design and evolution of the oscillatory response under light–dark cycles. *Biochimie, 90*(6), 888–897.

Rodrigo, G., Carrera, J., & Jaramillo, A. (2011). Computational design of synthetic regulatory networks from a genetic library to characterize the designability of dynamical behaviors. *Nucleic Acids Research, 39*(20), e138.

Rodrigo, G., & Jaramillo, A. (2008). Computational design of digital and memory biological devices. *Systems and Synthetic Biology, 1*(4), 183–195.

Romer, K., & Mattern, F. (2004). The design space of wireless sensor networks. *Wireless Communications, IEEE, 11*(6), 54–61.

Salis, H. M., Mirsky, E. A., & Voigt, C. A. (2009). Automated design of synthetic ribosome binding sites to control protein expression. *Nature Biotechnology, 27*(10), 946–950.

Sarle, W. S. (1994). Neural networks and statistical models. In *Nineteenth annual SAS users group international conference*, SAS Institute.

Schwille, P. (2011). Bottom-up synthetic biology: Engineering in a tinkerer's world. *Science*, *333*(6047), 1252–1254.

Searls, D. B., Guan, Y., Dunham, M., Caudy, A., & Troyanskaya, O. (2010). Systematic planning of genome-scale experiments in poorly studied species. *PLoS Computational Biology*, *6*(3), e1000698.

Sharp, P. M., & Li, W.-H. (1987). The codon adaptation index-a measure of directional synonymous codon usage bias, and its potential applications. *Nucleic Acids Research*, *15*(3), 1281–1295.

Shetty, R. P., Endy, D., & Knight, T. F. Jr, (2008). Engineering BioBrick vectors from BioBrick parts. *Journal of Biological Engineering*, *2*(1), 1–12.

Smith, B., Ashburner, M., Rosse, C., Bard, J., Bug, W., Ceusters, W., et al. (2007). The OBO Foundry: Coordinated evolution of ontologies to support biomedical data integration. *Nature Biotechnology*, *25*(11), 1251–1255.

Stricker, J., Cookson, S., Bennett, M. R., Mather, W. H., Tsimring, L. S., & Hasty, J. (2008). A fast, robust and tunable synthetic gene oscillator. *Nature*, *456*(7221), 516–519.

Suarez, M., Rodrigo, G., Carrera, J., & Jaramillo, A. (2010). Computational design in synthetic biology. In M. Schmidt, A. Kelle, A. Ganguli-Mitra & H. Vriend (Eds.), *Synthetic biology* (pp. 49–63). Netherlands: Springer.

Temme, K., Hill, R., Segall-Shapiro, T. H., Moser, F., & Voigt, C. A. (2012). Modular control of multiple pathways using engineered orthogonal T7 polymerases. *Nucleic Acids Research*, *40*(17), 8773–8781.

Thompson, A., Layzell, P., & Zebulum, R. (1999). Explorations in design space: Unconventional electronics design through artificial evolution. *IEEE Transactions on Evolutionary Computation*, *3*(3), 167–196.

Tigges, M., Marquez-Lago, T. T., Stelling, J., & Fussenegger, M. (2009). A tunable synthetic mammalian oscillator. *Nature*, *457*(7227), 309–312.

Tomita, M. (2001). Whole-cell simulation: A grand challenge of the 21st century. *Trends in Biotechnology*, *19*, 205–210.

Tomshine, J., & Kaznessis, Y. N. (2006). Optimization of a stochastically simulated gene network model via simulated annealing. *Biophysical Journal*, *91*(9), 3196–3205.

Torra, V. (2002). A review of the construction of hierarchical fuzzy systems. *International Journal of Intelligent Systems*, *17*(5), 531–543.

van Laarhoven, P. J. M., & Aarts, E. H. L. (1987). *Simulated annealing: Theory and applications.* Dordrecht: D. Reidel Publishing Company.

Wang, B., Kitney, R. I., Joly, N., & Buck, M. (2011). Engineering modular and orthogonal genetic logic gates for robust digital-like synthetic biology. *Nature Communications*, *2*, 508.

Weeding, E., Houle, J., & Kaznessis, Y. N. (2010). SynBioSS designer: A web-based tool for the automated generation of kinetic models for synthetic biological constructs. *Briefings in Bioinformatics*, *11*(4), 394–402.

Welch, M., Govindarajan, S., Ness, J. E., Villalobos, A., Gurney, A., Minshull, J., et al. (2009). Design parameters to control synthetic gene expression in *Escherichia coli. PLoS One*, *4*(9), e7002.

Wong, E., Baur, B., Quader, S., & Huang, C.-H. (2012). Biological network motif detection: Principles and practice. *Briefings in Bioinformatics*, *13*(2), 202–215.

Yang, X., Senthilkumar, D., Sun, Z., & Kurths, J. (2011). Key role of time-delay and connection topology in shaping the dynamics of noisy genetic regulatory networks. *Chaos: An Interdisciplinary, Journal of Nonlinear Science*, *21*(4), 047522–047526.

You, L., Cox, R. S., Weiss, R., & Arnold, F. H. (2004). Programmed population control by cell–cell communication and regulated killing. *Nature*, *428*(6985), 868–871.

Zhang, S., Wang, R.-S., & Zhang, X.-S. (2007). Uncovering fuzzy community structure in complex networks. *Physical Review E*, *76*(4), 046103.

Zhou, A., Qu, B.-Y., Li, H., Zhao, S.-Z., Suganthan, P. N., & Zhang, Q. (2011). Multiobjective evolutionary algorithms: A survey of the state of the art. *Swarm and Evolutionary Computation*, *1*(1), 32–49.

CHAPTER 2

Constraints in the Design of the Synthetic Bacterial Chassis

Antoine Danchin*,[†],[‡],[1], Agnieszka Sekowska*

*AMAbiotics, Evry, France; Department of Biochemistry, Faculty of Medicine, The University of Hong Kong, Hong Kong, and CEA/Genoscope, Evry, France
[†]Department of Biochemistry, Faculty of Medicine, The University of Hong Kong, 21 Sassoon Road, Pokfulam, Hong Kong
[‡]CEA/Genoscope, 2, rue Gaston Crémieux, Evry, France
[1]Corresponding author. e-mail address: antoine.danchin@normalesup.org

1 INTRODUCTION

The practical goal of synthetic biology (SB), putting aside the name, and its fundamental goals, consists in the construction of living cells from parts, which are assembled to produce a functional entity (Danielli, 1972, 1974). Implicit in this practice are both an engineering stance (i.e. conceiving an appliance and putting parts together in a rational, preferably standardised way) and the idea that one needs to split between a program (meant to put in effect the goal of the construct, usually production of some chemical compound or energy) and a machine (named *chassis*, plural chasses, by the geeks of the SB community) that expresses the program. The vast majority of the work of the SB community is to design new programs by assembling parts that may be reused for implementing further designs (Shetty, Endy, & Knight, 2008; Ellis, Adie, & Baldwin, 2011). Programs in SB terms are perceived as similar to electronic engineering constructs and depicted using the standard drawing of combinatorics of logical gates (Gendrault et al., 2012). In this context, extant living cells are seen as readily available multipurpose machines that can run the synthetic programs, assumed to have more or less universal portability (i.e. usability of the same construct in different environments) (Silva-Rocha et al., 2013). Yet, even in the computer industry, it is well known that portability is far from straightforward. Idiosyncrasies of the reading/running machine are essential in this respect, and they are the source of many bugs and safety loopholes. Furthermore, biologically engineered constructs are not necessarily analogous to electronic systems (e.g. no wires and no current), so that the analogy falls short of expectancies in terms of reproducibility (see how individual cells behave in the toggle-switch example (Elowitz & Leibler, 2000; Gardner, Cantor, & Collins, 2000)) and predictability. In what follows, we do not analyse the constraints that govern the program's chassis, as this is the focus of the vast majority of studies involving SB. As a matter of fact, most generally, terms such as 'biobricks' refer to standardised components of the genetic program, obliterating any consideration about the machine that is required to read it. However, to go for efficient SB constructs, we need to understand the constraints that govern the cell's behaviour (Danchin, 2012). Placing some of those in proper context is the purpose of the present chapter. To this aim, we use functional analysis (Cole, 1998; Fantoni, Apreda, & Bonaccorsi, 2009) and focus on the less obvious functions that have to be taken into account in the ideal chassis. In general, the references we propose below are from the recent literature, but we are careful to cite old work when there would be a danger (unfortunately becoming common today) of reinventing the wheel.

2 TOP-DOWN VERSUS BOTTOM-UP FRAMING

Functional analysis is an essential step in the design of industrial devices and processes (Cole, 1998; Fantoni et al., 2009). Two complementary approaches are used in this heuristic methodology. The first is a bottom-up approach, usually based on previous knowledge of systems similar to the object of interest. This approach begins with listing parts and proceeds by combining them into progressively more complex entities until the final instrument is obtained. The second is a top-down approach that starts from the end-user point of view, identifying first the *master function* of the device (e.g. moving on roads and streets for a car while transporting passengers and other loads) and then progressively going down to the *helper functions* required for the master function to operate (e.g. an engine, an energy store, lights, doors, windows, and seats for a car), to end up with the basic components making the device (Figure 2.1).

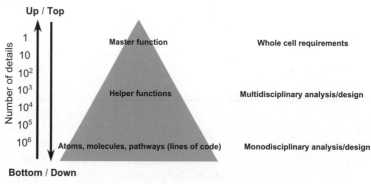

FIGURE 2.1

Functional analysis. The heuristics starts from identification of the master function of the contraption to be designed and processes by associating it with helper functions until the analysis identifies the nuts and bolts of the lowest level required for the building up of the designed object of interest. (For interpretation of the references to colour in this figure legend, the reader is referred to the online version of this chapter.)

The former approach has commonly been discussed with emphasis on the quest for the elusive minimal genome as the way to get access to all necessary components defining a living cell (Forster & Church, 2006; Juhas, Eberl, & Glass, 2011; Zhang, Chang, & Wang, 2011). Briefly, the implicit (rarely explicit) aim of that approach is to try and identify all the *functions* that are ubiquitously present in extant genomes. However, this approach makes use of the fact that genomics allows us to have access to ubiquitous *structures* (or, rather, gene *sequences*, which are seldom sufficient to have access to structures) and not functions. This view assumes that comparing genomes as languages carved on the Rosetta stone would lead to the Grail of a conserved minimal genome, and assuming that the gene tells the structure and the structure tells the function will allow us to reach the requested goal. Unfortunately, this is but a dream, as there is no unique correspondence between structure and function (Galperin, Walker, & Koonin, 1998; Danchin, 1999). The lack of this correspondence means that we must use heuristic methods to get to the point and either infer functions from structures (this is the common bottom-up practice) or design functions from an analysis of the goal to be achieved (a top-down approach). In fact, while there could well exist a minimal set of functions required for life to develop, there is no universal set of genes enabling these functions (Lagesen, Ussery, & Wassenaar, 2010). An efficient heuristic to overcome this problem is the identification of *persistent genes*, that is, genes that tend to be present in a quorum of genomes with a preset conservation percentage threshold (Fang, Rocha, & Danchin, 2005). This formulation has been reviewed by Acevedo-Rocha, Fang, Schmidt, Ussery, and Danchin (2013). This classification divides the genome into two components, the *paleome* that codes for all functions needed to reproduce the cell in its progeny, while replicating its genome, and the *cenome* that allows the cell to belong to a specific environmental niche. We shall not expand further this bottom-up approach that has been extensively discussed, as our aim is to use the engineer's top-down approach (what should not be forgotten in the design when constructing an effective and useful appliance?), to provide a rational set of functions needed to achieve SB goals. This approach has seldom been developed.

The top-down approach is key to the engineering practices used for the design of human artefacts. It begins with establishing the master function of the designed construct. Subsequently, it defines the helper functions that are needed to run the master function. Finally, it ends up in basic components, nuts and bolts, that are necessary to build up the system of interest. This approach is hierarchical, a constraint that is not as straightforward as it may seem to be. Indeed, there are two major types of hierarchies in living organisms that need to be understood and put together in the final construct. The hierarchy that is easiest to understand is the *dendritic hierarchy* that develops as a tree, combining the lowest-level parts into progressively higher-level entities. In the cell, atoms combine into molecules that form macromolecules and then complexes and then organise into a cell. In multicellular organisms, the body comprises several treelike helper function systems, such as the nervous system, the respiratory system, or the blood system. These systems obey rules that allow them to occupy as much space as possible, so that they can coordinate the whole behaviour of the organism.

In terms of physical organisation, it has been proposed that these systems tend to be scale-free or fractal (see, e.g. Sporns, 2006; Copley et al., 2012; Reese, Brzoska, Yott, & Kelleher, 2012). This view came from analysis of the networks of interactions explored at different scales via the 'guilt by association' inference. Guilt by association provides a general top-down principle for analysing gene networks in functional terms or assessing their quality in encoding functional information (Nitschke et al., 1998; Aravind, 2000). However, despite the popular success of the concepts of scale-free structures and fractals, the fact that interaction networks were scale-free is much disputed. For example, it was observed that the structure of protein–protein interaction networks is better represented by a geometric random graph than by a scale-free model (Przulj, Corneil, & Jurisica, 2004). And the very idea of association had to be qualified and split into different categories (proximity in the chromosome (Overbeek, Fonstein, D'Souza, Pusch, & Maltsev, 1999), in protein networks (Bachman, Venner, Lua, Erdin, & Lichtarge, 2012), in codon usage bias (Bailly-Bechet, Danchin, Iqbal, Marsili, & Vergassola, 2006), in amino acid composition (Pascal, Medigue, & Danchin, 2005; Heizer et al., 2006), in metabolic pathways (Nitschke et al., 1998), in concerted evolution (Pazos & Valencia, 2008; Engelen, Vallenet, Medigue, & Danchin, 2011), and even in the literature: genome analysis 'in biblio' (Nitschke et al., 1998)). It appears that multifunctionality, rather than association, is a primary driver of gene function prediction (Gillis & Pavlidis, 2011). Furthermore, gene function is not consistently encoded in networks, but rather depends on specific critical interactions. This implies that the details of how networks encode functions and what information computational analyses use to extract functional meaning are essential. As a consequence, the topic of SB quite similar to standard engineering practice, where 'the devil is in the details', superimposed on the general laws of physics and chemistry at work in living organisms. For example, it has been shown that the network structure itself provides clues as to which connections are critical and which systemic properties, such as scale-free-like behaviour, do not map onto the functional connectivity within networks (Gillis & Pavlidis, 2012). In short, as in all engineering practices, SB has to take details into account, even when they look anecdotal. We shall go back to this point in the succeeding text when discussing metabolic frustration. We might notice here that this concept fits quite well with the 'tinkering' ('bricolage') properties of life (Jacob, 1974), suggesting that the attitude of the engineer is well adapted to understand what life is.

Besides dendritic hierarchies, living organisms display a *segmented hierarchy*, that which is found in animals, where organs play a central role, complementary to that of the general connecting systems forming a dendritic hierarchy. This type of hierarchy is found not only in Eukarya, with their variety of organelles, but also in Bacteria, where many microcompartments exist (such as acidocalcisomes (Seufferheld, Kim, Whitfield, Valerio, & Caetano-Anolles, 2011), carboxysomes (Bonacci et al., 2012), magnetosomes (Greene & Komeili, 2012), and many other structures as well (Souza, 2012)). Analysis of compartmentalisation, with emphasis on the relationships between these hierarchies, will therefore be an essential step in SB-related functional analysis. This is a complex area that will not be further developed here.

3 AN UNLIMITED LIST OF FUNCTIONS (ORGANISED STARTING FROM THE CELL'S STRUCTURE)

The first function that must be defined is the master function of the SB appliance we wish to construct. The identification of this function is by no means trivial as SB-related master functions fall into two broad categories, depending on the aim of the project that is undertaken. Most often, the cell (or the organism) is seen as a factory. The master function of a factory is its output, which may correspond to a specific entity or related families of entities. With this view, the growth of the cell population is an associated helper function, not the master function. In contrast, when SB is meant to help investigators to understand what life is, the master function of the cell coincides with life itself, where making a progeny and growth is essential. Two main functions are generally ascribed to life, making a progeny (Jacob, 1974) and, via pre-eminence of movement, as in the long tradition following Aristotle, exploring the environment (Danchin, 2003). Interestingly, these functions correspond to the genome split: the paleome drives construction of a progeny, while the cenome allows populations of cells to occupy a specific environmental niche (Acevedo-Rocha et al., 2013). In what follows, in order to avoid dwelling into the deep philosophical questions raised by the nature of what life is, we chose to limit our discussion to the first family of functions, considering the cell as a factory meant to produce specific outputs and considering the general properties of life as the associated helper functions. This choice fits with the standard view of SB, where the construction of goal-directed programs is the favoured engineering procedure.

To this aim, we consider the cell factory as a whole, and progressively focus on its internal components, while identifying processes that are important in some aspects of SB development. We follow a scenario in which life is perceived as the process that would allow computers to make computers ((Brenner, 2012; Danchin, 2009b) and see the RepRap project for a first step in this direction http://www.reprap.org/). This scenario combines three major helper functions: compartmentalisation (defining an inside and an outside, and shaping), information transfer (defining a program, with relevant hardware to support the corresponding memory and memory transfers), and metabolism (the process that selects basic building blocks from the environment, constructs the cell's components, and manages energy, releasing waste in the environment).

3.1 Compartmentalisation and shaping: the cell's casing
3.1.1 Interface with the environment
The cell interacts with its environment in a selective way, with interactions being mediated by the cell envelope. Moreover, the envelope is not only involved in transport but is also required for other functions, such as protection.

Protection is important, as most biological compounds are fragile, and this evolved in a huge variety of envelope structures and protective processes that need to be understood and implemented in SB applications.

Physical protection: Cells only bordered with a lipid bilayer seeded with proteins are rare and thrive only in highly protected niches (such as the surface of mucosa). Bacteria have evolved a variety of protective structures, including outer membranes (Page, 2012) found in diderms (see Gupta (2011), with Mycobacteria as a recently recognized addition (Stoop, Bitter, & van der Sar, 2012)), simple or multiple layers of peptidoglycan (Callewaert et al., 2012), teichoic acids (Hanson & Neely, 2012), and S-layer proteins (Pavkov-Keller, Howorka, & Keller, 2011). Archaea have evolved special lipids (Chong, Ayesa, Daswani, & Hur, 2012). Eukarya have many special protective structures, including intercellular compounds such as the plant cell wall, made of cellulose, lignin, and pectin (Gibson, 2012).
Chemical protection: Cells may have to compete with other organisms. As protective means, they developed a wealth of not only toxic compound (e.g. colicins (Cascales et al., 2007) and toxins (Norton & Mulvey, 2012) but also small molecules such as antibiotics (Davies & Davies, 2010)).
This type of protection is coupled to transport, which we consider now.

Transport is essential both ways (import and export) and it is at the root of the electrochemical potential that allows the cell to extract specific compounds from its environment and produce energy.

Electrochemical potential: Ion selectivity is central to establish an active electric potential between the inside of the cell and its outside, and this difference is essential for energising the cell. Potassium is ubiquitous inside living cells, despite the fact that it is far less abundant in the environment than sodium. This implies the existence of selective transport systems that discriminate between similar ions (Corry & Thomas, 2012). The details of the discrimination process illustrate vividly the challenges faced by the SB engineer when they have to choose or construct a cell chassis. For example, in an ion permease, positional constraints on the ion-coordinating atoms create a preference for binding one ion type over another (cavity effects). Three types of effects, the 'rigid cavity', the 'strained cavity', and the 'reduced ligand fluctuation' (RLF) mechanisms, have been proposed to explain ion selectivity in typically engineering terms, as follows (discussed in Thomas, Jayatilaka, and Corry (2013)). In the *rigid cavity* model, ion selectivity is due to the rigid backbone of the permease, regardless of the type of ion. For example, some positions will energetically favour Na^+ over K^+, creating selectivity for that ion. However, thermal motion is unavoidable. It is often larger than the size difference between Na^+ and K^+. Nevertheless, if the ligands fluctuate about some fixed average position for different ions, this will still create ion selectivity, yet the ligands are likely to fluctuate about different average positions when coordinating ions of different size, and the difference in average ion-ligand distance when coordinating Na^+ and K^+ is almost always similar to the difference in ion size, which argues against a true rigid cavity as a common structure. Another model, similar to the induced-fit model (Johnson, 2008), has therefore been proposed: a *strained cavity* allows for the average ion–ligand distances to adjust according to different ion types. However, the adjustment comes at a cost that needs to be accounted for. Furthermore, this situation is unlikely to be widespread, due to the inherent flexibility of proteins. Using simple abstract-ligand models, Thomas and coworkers noted that cavity-based mechanisms could still create selectivity even both when the cavity size is not fixed and when there is no strain associated with adjustment of the ligand positions. In this case, the only cavity factor controlling the binding energy of the ions was the degree of thermal motion associated with the ligands as the ligands were free to adopt their optimum positions for each ion type. This is the essence of the entropy-driven RLF model for ion selectivity (Thomas et al., 2013). The elucidation of this mechanism offers a more complete picture of the ways in which the fundamental process of ion selectivity can be achieved. Engineering details will have to be understood to allow SB constructs to coax ion transport into efficient design. Some cells have already been identified that generate electric current (Desloover, Arends, Hennebel, & Rabaey, 2012; Khunjar, Sahin, West, Chandran, & Banta, 2012; Fu et al., 2013) and that form nanowires (Pfeffer et al., 2012), features that will probably be at the root of interesting engineering developments.

Safety valves: In addition to the building up of a significant electrochemical potential, the cell has to cope with considerable variations in osmolarity. For example, in the infant mammalian gut, milk-feeding induces the lactose operon in *Escherichia coli* cells. At the time of the next feed, the cells experience suddenly a surge of the sugar. Presence of preinduced lactose permease (LacY) in the cell's membrane will trigger a rapid influx of lactose. Because LacY allows the building up of steep gradients, the intracellular concentration of lactose may reach a level that raises the osmotic pressure so much that the cells will burst or at least leak ((Dykhuizen & Hartl, 1978), and this gives a hand for the selection of *lacY*-defective mutants (Hopkins, 1974; Varela, Brooker, & Wilson, 1997)), unless some protective mechanism exists. A safety valve of the type that engineers place on steam engines to cope with excess pressure is a suitable answer. A natural way to engineer the process would be to modify lactose into a component that cannot use LacY and then to excrete the modified compound, which would be unable to get back into the cell. The first half of the process fits exactly with the function of lactose acetyltransferase, coded by *lacA*, the third gene of the lactose operon, including with its surprisingly poor K_M (in the molar range), now explained as its contributes to fitness via making acetyl-galactosides only when the lactose concentration is high, so that normal amounts of lactose can still be used by the cell (Danchin, 2009c). In this case, modified carbohydrate permeases such as SetA, SetB, or SetC (Maa for maltose derivatives) play the role of safety valves in *E. coli*, exporting the modified sugar. Counterparts exist in most bacterial clades, and acetyltransferases are ubiquitously present (while generally poorly annotated). Engineering trios that link import, modification, and export will be of major importance in all SB constructs that aim at producing large amounts of chemicals.

Storage: An alternative to safety valves is the storage of metabolites in the form of polymers, for example, to cope with high osmolarity. Glycogen (Chandra, Chater, & Bornemann, 2011), polyhydroxyalkanoates (Guerrero & Berlanga, 2007), and polyphosphate (Achbergerova & Nahalka, 2011) are found in many cells (the latter is ubiquitous) and this fills two functions, buffering osmotic pressure and storing important compounds for situations where nutrients and energy are limiting.

3.1.2 The need for a casing or a skeleton

Cells build up a considerable osmotic pressure. This implies that, in the absence of a particular casing or skeleton, the cell will be spherical in shape. Protecting devices such as the cell wall will result in providing cells with a variety of shapes, usually that of bacilli. However, many other more or less complex forms, including helical structures, can be observed (see Bergey's Manual for a myriad of examples (Bergey's Manual Trust, 1986)). Tamames and coworkers explored the bold hypothesis that there might be a relationship between the structure of the bacterial chromosome and the cell's shape, involving genes that are somehow shaping the cell. Analysing the *mur–fts* clusters that exist in bacteria with a cell wall and have a quite variable distribution in the genome, they uncovered an unexpected pattern of relationships between the order of the genes in the chromosome and the shape of the bacteria (Tamames, Gonzalez-Moreno, Mingorance, Valencia, & Vicente, 2001). Remarkably, ranking bacteria by shape, the tree of shapes did not match the phylogenetic tree, but that of the organisation of the *mur–fts* clusters, suggesting a deep relationship between the order of these genes and the architecture of the cell. The authors later proposed a model in which the selective pressure to maintain the division cell wall cluster arises from the need to efficiently coordinate the processes of elongation and septation in rod-shaped bacteria (Mingorance, Tamames, & Vicente, 2004). Indeed, the asymmetry of the cell's volume in bacilli-shaped bacteria accommodates entropy-driven chromosomal segregation (Danchin, Guerdoux-Jamet, Moszer, & Nitschké, 2000; Jun & Mulder, 2006). This is not straightforward in cocci, as symmetry breaking will be needed to allow unambiguous split of chromosomes into daughter cells (Harold, 2007). This shaping constraint needs to be explored when choosing a chassis for SB. Experiments to validate Tamames and coworkers' observation (e.g. by manipulating the gene order in *mur–fts* clusters) are necessary to implement the corresponding constraints in a rational manner.

During growth, the existence of a casing, as illustrated nicely in insects or crustacea, poses specific problems that are solved in a variety of ways, such as moulding. The complex organisation of the cell division machinery is the counterpart for unicellular organisms. An alternative, usually complementary solution is that of a skeleton that can grow in parallel with the organism for a significant amount of time. Bacteria have structures that play the role of a skeleton, with FtsZ and MreB as homologues of tubulin and actin, respectively (Celler, Koning, Koster, & van Wezel, 2013; Pilhofer & Jensen, 2013; Wang & Shaevitz, 2013). At this point, it is not clear whether the cytoskeleton drives the shape of the cell wall or whether the cytoskeleton is constrained in its shape by the bacterial wall (Wang & Wingreen, 2013). A question of major importance here is the involvement of energy in the shaping/division process, as it may have a significant informational component (Binder & Danchin, 2011) that needs to be taken into account in any SB design of a chassis.

3.2 Information transfer

While information is already present in the machine (as it is in the computer that runs the information of its operating system), the way most investigators use the category information in SB today is restricted to the genetic program and its expression (Endy, 2005). We chose here to focus not on the SB-specific genetic programs, but on the constraints of the machineries required for information transfer. Two types of chassis-related information transfer must be taken into account to deal with the genetic program: replication (i.e. exact duplication) of the program and expression of the program (depending on environmental conditions, and essential to the cell factory goal, and comprising transcription and translation). In both cases, a sequence of specific functions is required: initiation of the process, elongation, and termination, in parallel with proofreading. In addition, it is important that maintenance functions are implemented to keep working the various avatars of the program.

3.2.1 Synthesis of macromolecules

Both DNA replication and gene transcription require nanomachines that break the DNA double helix open, bind into the opened structure, and start elongating a nucleic acid molecule in the $5'-3'$ direction. The process is relatively straightforward in transcription and in replication of the *leading* DNA strand, so much so that it has already been included in xenobiological constructs, involving analogues of the natural nucleotides ((Giraut, Abu El-Asrar, Marliere, Delarue, & Herdewijn, 2012). Even viruses can use nonstandard nucleotides and 2-amino-adenine also exists in extant cyanophages (Khudyakov, Kirnos, Alexandrushkina, & Vanyushin, 1978)). However, when engineering replication, it is important to refrain from using energy to unfold the double helical structures (because the cost of replicating the program would then be far too high), while the use of specific, possibly energy-dependent contraptions to manage supercoiling could be allowed (Pommier, 2013). The former constraint is driven by helicases that use entropy-driven processes (no free energy, therefore, as proven long ago by Lifson and Zimm, 1963) to propagate the transcription or replication bubble. By contrast, other energy-dependent helicases such as those involved in repair or recombination (Smith, 2012) use free energy when they are involved in specific information-managing functions (Danchin, 2009a). The solution for replicating long DNA segments must be completely different, however, as replication of the *lagging* strand poses major structural problems. Indeed, replication of that strand requires a considerable length of single-stranded DNA that must be protected by specific complexes; it also requires management of multiple initiation complexes, in contrast to replication of the leading strand (Balakrishnan & Bambara, 2013).

Following transcription, the protein biosynthetic machinery brings complexes composed of ribosomes, chaperones, and localisation factors into similar actions (begin, elongate, and end). This machinery also interacts directly with factors dedicated to disposal of protein fragments (generated during mistranslation, translation interruption, or premature termination) and more generally to protein degradation (Rodrigo-Brenni & Hegde, 2012; see also the succeeding text).

Many further functions must be considered in the manufacture of macromolecules and eventually implemented in SB constructs. Most of these considerations address the fact that the threadwire machinery that makes macromolecules cannot fold them readily into their final proper three-dimensional (3D) shape (discussed in (Danchin, 2012). These factors also account for the difficulties encountered with genome transplantation (see also the succeeding text, the short discussion of the fourth dimension) and in maintenance of cell shape. Some of these factors are proposed here.

Scaffold: As in the construction of large buildings, large or fragile structures cannot reach their final form immediately. They need temporary scaffolding (Saberi & Emberly, 2010; Tamaru et al., 2011; Wirth et al., 2012). In some cases, scaffolding is associated with a further function, that of measuring length (the scaffold's subunits within a virus tail do not have the same helical pitch as the tail subunits, which can be used vernier scale-like (Ackermann, 1998)). Scaffolds may also be associated to the energy-dependent creation of a young progeny, distributing aged components in specific compartments or locations, Maxwell's demon-like (Binder & Danchin, 2011; Danchin, 2012).

Shape: Achieving the final shape of a protein often needs helper functions at a lower hierarchical level. This function is at the root of the concept of molecular chaperones, proteins (or possibly RNAs (Grohman et al., 2013)) that shape relevant substrates while protecting them against misleading interactions. The history of this concept can now be traced back fairly well. The first time when the term chaperone was published is in a work by Laskey where he showed that the nucleosome subunits of chromatin are assembled from histones and DNA by an acidic protein that binds histones in a nonstoichiometric way. This nucleosome assembly protein was identified and purified from eggs of *Xenopus laevis*: "we suggest that the role of the protein we have purified is that the two subunits of a 'molecular chaperone' which prevents incorrect ionic interactions between histones and DNA" (Laskey, Honda, Mills, & Finch, 1978). Later on, Laskey coined the name nucleoplasmin for the protein. In a parallel and independent work on the plant enzyme RuBisCO, Ellis had discovered a factor likely to promote assembly of the mature enzyme. At some point, becoming aware of the previous work of Laskey (not immediately coming to his mind because in very different domains of biology, such as plant and animal biology, scientists do not mix up very much), Ellis wondered whether this could not be the hallmark of a general function. In 1985, at a meeting of the Royal Society, he proposed a novel concept for informational macromolecules, that of proteins involved in shaping other proteins. Two years later, Ellis summarised the views he had presented at another meeting held in Copenhagen: "At a recent meeting I proposed the term 'molecular chaperone' to describe a class of cellular proteins whose function is to ensure that the folding of certain other polypeptide chains and their assembly into oligomeric structures occur correctly" (Ellis, 1987). With these two authors, the concept of molecular chaperone was born. No more inventive activity was necessary to make obvious that it would be interesting to explore further generalisations. Indeed, it was rapidly recognised by many authors that the corresponding function was ubiquitous and that as a matter of fact it had already been unknowingly witnessed in the bacterial proteins allowing phage morphogenesis (the GroESL system) and involved in the heat-shock response (Tilly & Georgopoulos, 1982). Counterparts could be found in most, if not all, organisms (Bogumil & Dagan, 2012). It was also observed that their activity could be regulated by posttranslational modifications (Cloutier & Coulombe, 2013). Proper folding is intimately linked to proper function, and chaperones are therefore involved also in quality control and maintenance. With SB engineering in mind, there is still much to be understood about the corresponding shaping function as, while some molecular chaperones use energy (da Silva & Borges, 2010), some, such as the trigger factor or peptidyl prolyl isomerases, do not (Hoffmann, Bukau, & Kramer, 2010; Preissler & Deuerling, 2012; Schiene-Fischer, Aumuller, & Fischer, 2013), pointing out a considerable difference in their role, not yet emphasised but certainly of major importance for engineering. Finally, the role of RNA chaperones begins only to be understood (Grohman et al., 2013).

Mould: Understanding protein and RNA folding remains difficult. While it is accepted that it is an entropy-driven process where water plays an essential role, the actual minima reached as the final structures are still difficult or even impossible to predict. The idea of constraints due to structural stability (a concept central in the views of the mathematician René Thom (Thom, 1989)) established that folding may occur via scale-free constraints (Petersen, Neves-Petersen, Henriksen, Mortensen, & Geertz-Hansen, 2012), but we have seen earlier that this fashionable view may be misleading. In the case of proteins, the process of allostery illustrates that several different conformations may be simultaneously present, a property widely exploited for regulation (Tzeng & Kalodimos, 2011; Nussinov, Tsai, & Ma, 2013). The discovery of allosteric transitions extended a major chemical question, how do proteins take their shape, and where do they get the necessary information? Proteins cannot suddenly reach their final shape. Molecular chaperones act as guides, but is this the whole story? Is there a specific feature that points out where information (literally 'giving form') comes in? Citing Motlagh, Li, Thompson, and Hilser (2012): "Allostery is a biological phenomenon of critical importance in metabolic regulation and cell signalling. The fundamental premise of classical models that describe allostery is that structure mediates 'action at a distance'. Recently, this paradigm has been challenged by the enrichment of IDPs (intrinsically disordered proteins) or ID (intrinsically disordered) segments in transcription factors and signalling pathways of higher organisms, where an allosteric response from external signals is requisite for regulated function. This observation strongly suggests that IDPs elicit the capacity for finely tunable allosteric regulation", we see that protein 'order' and allostery go hand in hand. This hints at a moulding function for the disordered proteins and disordered regions ('unstructured' or 'flexible' may be preferable) in proteins proposed by Dunker and colleagues at the end of the 1990s (Dunker et al., 1998; Romero et al., 1998). At the time, this view had not been included in the many reflections of specialists of protein folding, spanning over several decades, who tried to understand the shaping of proteins. It was therefore a challenging idea that took some time to be accepted. Subsequently, ideas tend to come into clusters, several investigators embarked on the study of the consequences of this discovery, and it still took some time for the international community to accept this view, which is now well recognised and universally used: even ribosomal proteins have flexible regions (Timsit, Allemand, Chiaruttini, & Springer, 2006). A major role of such flexible proteins is to mould on their substrate or make a collaborative interaction that results in the final shape. This role (further discussed in the succeeding text, when we explore the fourth dimension) involves time in a way that will be essential for implementing functionally optimised SB constructs.

Proofread: As in standard engineering processes, macromolecule synthesis involves many steps. Each step is prone to go astray because of thermal noise, variation in the availability of required components, or interference by other competing processes. A rule of thumb shows that the average error rate per step is of the order of 10^{-3}–10^{-4} (Yadavalli & Ibba, 2012). The loading of an amino acid on its cognate tRNA is one such step. Several chemical processes prevent wrong tRNA acylation, via deacylation or transacylation, for example (Freist, Sternbach, Pardowitz, & Cramer, 1998). The selection of the cognate codon–anticodon interaction in the ribosome is a further hurdle, which is solved by the use of GTP-dependent elongation factors EFTu/EFTs. To keep the error rate low implies steps for efficient error correction (Hopfield, 1974; Thompson, 1988). Yet the final outcome still results in the synthesis of proteins having, on average, an error rate of the order of 10^{-3}, showing that every single protein deviates from the ideal sequence by at least one residue, a feature that is important for SB as this differs from most standard engineering processes where such a variation would not be tolerated. A similar rule applies to replication and transcription. Proofreading processes correct errors on the fly, during the very process of macromolecule biosynthesis, or at a later stage, as seen in the next paragraph. Such processes, and possibly novel ones still to be discovered (involving postsynthesis processing of the macromolecule), are essential for optimising and scaling up SB production.

Repair: The final error rate in replication is remarkably low, of the order of 10^{-9} or lower, which implies superimposition of several types of proofreading (Fijalkowska, Schaaper, & Jonczyk, 2012), including repair processes. Sustained proofreading during polymerisation cannot correct every error. Furthermore, the inevitable ageing of macromolecules results in nonfunctional entities. Some of these errors may be repaired by specific processes (e.g. mismatch repair copes with remaining errors after replication (Fukui, 2010; Marinus & Casadesus, 2009)), and it is often assumed that repair is one of the role of molecular chaperones

(Maisonneuve, Fraysse, Moinier, & Dukan, 2008). In contrast with nucleic acids, proteins age rapidly, even in the absence of external insults. Flexibility favours cyclisation into L-succinimide of aspartate and asparagine residues of the protein backbone. This process is fast and frequent, in particular at asparagine–glycine (AsnGly) motifs (Robinson & Robinson, 2004), with concomitant deamidation of asparagine. Deamidation is a common stumbling block heterologous protein expression (Heukeshoven, Marz, Warnecke, Deppert, & Tolstonog, 2012; Shimura et al., 2013). Subsequently, L-succinimide hydrolyses spontaneously into L-isoaspartate and, if methylated (see hereafter), back into L-aspartate (in one-third of cases), yielding again L-isoaspartate in two-thirds of the cases. Later on, L-succinimide may even isomerise at a slow rate into D-succinimide and then lead to formation of D-isoaspartate and D-aspartate residues, with considerable modification of the structure and activity of the protein (Figure 2.2). As a consequence, proteins with flexible regions containing aspartate or asparagine residues are prone to change over time, leading to multiple states, depending on their physicochemical environment. These changes are likely to be programmed and the result of natural selection. For example, the large beta subunit of bacterial RNA polymerase contains conserved AsnGly motifs.

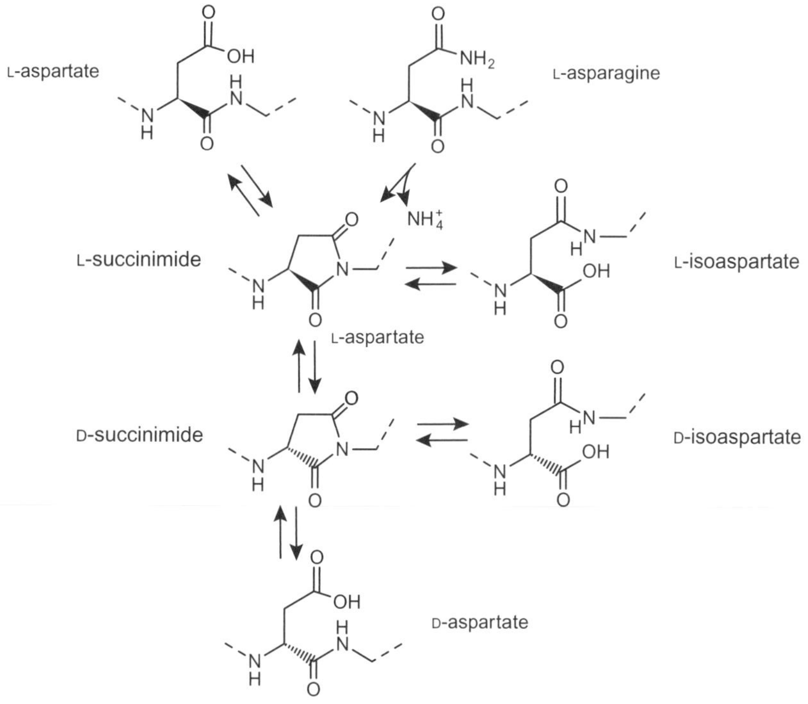

FIGURE 2.2

Spontaneous cyclisation and evolution of aspartate and asparagine residues in proteins.

From Danchin, Binder, and Noria (2011).

Because of the cost of resynthesising the protein, if doomed to become rapidly nonfunctional after cyclisation–deamidation, these motifs should have been genetically rejected rather than preferred, unless deamidation is of positive biological value (Danchin et al., 2011). In the same way, aspartate residues modified by phosphorylation or methylation are also prone to cyclisation. This may lead to loss of activity and creation of novel functional properties, such as regulatory properties or novel catalytic features (reviewed in Robinson and Robinson (2004)). SB engineering needs to take these constraints into account.

Distribute: Finding the proper place for a particular entity is very important for optimisation of SB processes. Most proteins have multiple contact regions that allow them to form connected complexes (Zhao, Hoi, Wong, Hamp, & Li, 2012). We have also seen that many proteins are partially unfolded. Unfolded regions may be used to create interactions that take the corresponding proteins to proper locations. This feature can be used in specific engineering constructs to target proteins at designed places in the cell (Barth, Schoeffler, & Alber, 2008). Finally, life has evolved a general process for the creation of novel proteins that stabilise complexes via the natural tendency of aromatic residues to make orthogonal interactions ('gluons') (Pascal et al., 2005). This particular trapping of information via enrichment of A+T-rich sequences may also be used in SB constructs, in particular for those that are obtained after controlled evolution processes (de Crecy-Lagard, Bellalou, Mutzel, & Marliere, 2001). The distribution of the genes in the genome has an effect in the final localisation of their products, as discussed. It is therefore important to orient genes correctly with respect to the movement of the replication fork (Rocha & Danchin, 2003), to place them at positions with similar codon usage bias (Bailly-Bechet et al., 2006), and to optimise the gene dosage (near or far from the origin of replication, the use of insertion spacers, etc. (Zucca, Pasotti, Mazzini, De Angelis, & Magni, 2012)) (Rocha, Guerdoux-Jamet, Moszer, Viari, & Danchin, 2000; Willenbrock & Ussery, 2004). As discussed, proteins age rapidly. While this may sometimes be part of their normal function (e.g. an aspartate residue from protein S11 from the *E. coli* ribosome is isomerised as isoaspartate within two minutes, hardly an unwanted property (David, Keener, & Aswad, 1999)), this may also be the first sign of deleterious senescence. Distribution of these modified proteins at special location in the cell (such as the cell poles or in a mother cell) may be important.

Degrade: In many situations, a malfunctioning entity cannot be repaired. Also, some components of the cell must have a limited lifespan. They must either be expelled out of the cell (which is not a convenient solution for large objects) or degraded into reusable components or components that can be exported. A specific complex, the degradosome, functionally but not structurally conserved in different bacterial clades (Rauhut & Klug, 1999; Danchin, 2009a; Lehnik-Habrink et al., 2010; Redko et al., 2013), and in Eukarya (the exosome (Chlebowski, Lubas, Jensen, & Dziembowski, 2013)), takes care of RNA turnover and degradation. This complex not only disposes of RNA but also couples degradation to anabolism via association with enzymes such as enolase and polynucleotide phosphorylase (see Nitschke et al. (1998) and Section 3.3, in the succeeding text). Currently, the role of degradation in protein turnover is much less clear. In contrast with many RNA molecules, most proteins must be stable. Nondiscriminating proteases are expected to be highly toxic for the cell because they would degrade functional and nonfunctional proteins equally well. Therefore, these kind of proteases are usually exported enzymes that use polypeptides as substrates. The most important intracellular proteases are energy-dependent enzymes. This is an unexpected feature because proteolysis is exothermic. It has been speculated that when energy is not used as expected, it is used for discrimination (Hopfield, 1974). Among the many ways allowing discrimination between functional and dysfunctional entities, energy may be used to *prevent* degradation of what is functional. This idea is at the root of the concept of Maxwell's demon, a thought experiment that uses energy to reset a selection system by erasing the memory of its past actions (Maxwell, 1871; Binder & Danchin, 2011; Danchin, 2012). In terms of their mode of action, many types of proteases may exist (including processive proteases). As in the case of RNases, the activity of proteases leads to leftovers that must be dealt with by specific degradation systems. While these processes have been fairly well described for very short RNA molecules (nanoRNAs (Liu et al., 2012)), much remains to be described for the fate of peptides. It seems likely that many genes of unknown or poorly known function will belong to this family of enzymes that degrade leftovers.

Trash: Maxwell's demons such as septins act as sorting machines (see references in Binder and Danchin (2011)), with the mother cell collecting aged proteins. This processes remind us of a widely spread cellular function, that of the garbage bin. Formation of inert aggregates is probably a ubiquitous way to dispose of useless or even deleterious compounds. Bacteria tend to place these aggregates at the cell poles, in particular older poles (Lloyd-Price et al., 2012). This process has the consequence that bacteria age, go senescent, and die (Stewart, Madden, Paul, and Taddei (2005), but see Wang et al., (2010) for a view challenging the results of the previous reference), death being the ultimate trashing function. This unavoidable constraint needs to be seriously taken into account in SB designs.

3.2.2 Regulation
The control of metabolic and development processes is an essential function of information transfer. Indeed, regulation lies at the core of the SB activities centred on the genetic program, and the bulk of the work dealing with biobricks and similar constructs is aimed at constructing sophisticated regulatory devices (Wang & Buck, 2012; Berthoumieux et al., 2013). This will not be explored further here as regulation is the focus of the vast majority of SB work (Ang & McMillen, 2013). Instead, some recent complements that may be of interest are listed that have not yet been widely discussed.

Connect: The objects that make a cell are connected together, forming networks allowing the analysis of the structure of networks identified as network regulatory motifs (Riccione, Smith, Lee, & You, 2013), an approach that has been described previously (Shen-Orr, Milo, Mangan, & Alon, 2002).

Control switches: Regulation deals with the control of central functions such as

begin (in particular attenuation of transcription via riboswitches (Wittmann & Suess, 2012) and regulation of translation initiation),
end (a neglected control stage, in particular at the level of translation, where UTP might couple transcription and translation termination (Noria & Danchin, 2002)).

Speed control: Acts on elongation (e.g. smooth regulation of translation speed possibly via a moderator tRNA form (Danchin & Dondon, 1979)); targeted DNA recognition complexes that can specifically interfere with transcriptional elongation, RNA polymerase binding, or transcription factor binding (Qi et al., 2013).

Sense: While the objects belonging to the cell are involved in a variety of pathways, the very fact that they are present or absent is an information that can be used by the cell to regulate a variety of processes. Sensing is therefore an essential function, widespread in cells, within many different relays. We have previously discussed allostery that modulates enzyme activities and other functions. Bacteria have often two-component systems made of a sensor that connects to a regulator usually via phosphorylation transfer, to control downstream processes (Kenney, 2010; Huynh & Stewart, 2011; Jung, Fried, Behr, & Heermann, 2012).

Stress gauge: In all cases when shapes have to change in a concerted fashion, a general function to sense physical stress is required. This function is the equivalent to what is known as *strain gauges* in engineering (Haswell, Phillips, & Rees, 2011; Mammoto, Mammoto, & Ingber, 2012). These gauges are meant to measure mechanical stress and transmit the information to relevant effectors. In general, all stresses can be used as indicators of the presence of information that must be taken into account by the cell's regulatory power. This implies comparing a state at the present time to reference states and therefore involves memory.

3.2.3 Memory

Finally, the management of memory is central for information transfer. Cells manage two major types of memory, genetic memory and epigenetic memory, processes that match the replication/reproduction dichotomy, a central tenet of the generation of a progeny. Cells reproduce (i.e. make similar copies of themselves), while the genetic program replicates (makes identical copies), and this matches with two time scales, two types of information (information of the chassis and information of the program), and two types of memories (Danchin, 2012).

Genetic memory evolves slowly across many generations as cells multiply, in parallel with a variety of proofreading functions, as discussed previously. Despite its slow rate changes, SB constructs cannot avoid taking genetic memory into account, especially when cells are involved in large-scale processes (e.g. scaling up will be essential for biofuel production). There is an inevitable accumulation of program variations as the number of generations increases. Cells behave as information traps (Danchin, 2009d) and this implies that some of these variations will improve the overall behaviour of the cell (we must note, however, that there is no in-built reason that this improvement would fit with the human design goals). Some do not, and the existence of sex, a function specific to biology, has been established as a process allowing to reset memory to its original state while propagating innovation (Hartfield & Keightley, 2012). Other processes allowing horizontal gene transfer will modulate genetic memory in ways that could be used for SB particularly for engineering large cell populations.

Epigenetic memory is managed at many levels: inter alia, tagging of DNA by methylation, positive autoregulation of membrane proteins synthesis, protein post-translational modifications, and protein spontaneous ageing/maturation (as discussed previously). Most of these modifications correspond to functions that have not yet been investigated in-depth, and it is likely that they will have presently unforeseen consequences in SB constructs. Hence, they should be monitored carefully. Finally, memory is not simply implemented in individual cells, but in cell populations, including formation of multicellular organisms, even in bacteria (Singh & Montgomery, 2011). The overall 'antifragile' behaviour of living organisms is tightly associated to populations (Chuang, 2012). Among those, populations made of cells of different species will most probably play a considerable role in the future of SB.

3.3 Metabolism

Small molecules involved in intermediary metabolism make up the core processes that are implicit in the functioning of the cell. These molecules do not exist as a haphazard collection of chemicals, but rather, their presence follows a chemically logical organisation. The logic of metabolism (not discussed in-depth here) implies the existence of specific constraints on the cell's building blocks: 20 (up to 22, with selenocysteine and pyrrolysine) proteinogenic amino acids, five nucleic bases, a series of coenzymes or prosthetic groups, some ubiquitous (pyridoxal phosphate, thiamine diphosphate, iron–sulphur clusters, heme, molybdopterin, etc.), lipids, and a variety of intermediary compounds that make the link between the different classes of essential metabolites. Catabolism aims at reducing available compounds to those basic building blocks while generating energy and disposing of toxic derivatives. Several functions are thus associated to the logic of metabolism.

Channel: Almost by definition, the chemical compounds that are present in the cell are reactive molecules. Most are involved in well-controlled metabolic pathways, but some may yield unwanted side reactions and this leads to a variety of behaviours that result in a variety of compartmentalisation processes (metabolic *frustration*, to use a term familiar to physicists to refer to organisation of multiple mutually incompatible entities (Hegler, Weinkam, & Wolynes, 2008)). This is illustrated, for example, by alpha-dicarbonyls (Kroh, Fiedler, & Wagner, 2008), reactive intermediates that pervade intermediary metabolism. The spatial optimisation via channelling from one catalytic site to the next one in the pathway where the molecule is used is a way to guarantee that such molecules do not go astray, and this property is indeed already used in SB (Agapakis, Boyle, & Silver, 2012).

Inactivate: The presence of some highly reactive intermediate metabolites in commonly used pathways is unavoidable. This is the case for many of the derivatives from pyridoxal-phosphate catalytic activity, which produces, in particular, highly reactive 2-aminoacrylate or related compounds. Enzymes belonging to the ubiquitous YjgF/YabJ/UK114 family (now RidA) inactivate these compounds into the corresponding keto acids (Lambrecht, Schmitz, & Downs, 2013). In the same way, enzymes of the nitrilase family will deamidate intermediates such as ketoglutaramate (Belda et al., 2013). Finally, it is likely that reactions assumed to be 'spontaneous' need in fact to be catalysed so that the corresponding reactive intermediate is inactivated.

Protect/Unprotect: In the chemist's laboratory, group protection is standard practice when a chemical reaction would tend to attack simultaneously several positions in a molecule while only one is the target. The same situation happens repeatedly in metabolism. For example, amino acids that differ from the twenty proteinogenic amino acids might be loaded onto tRNAs and subsequently be incorporated in proteins, altering their shape and function. Protection is the way out: non-proteinogenic amino acids are often N-acylated (acetylated or succinylated, depending on the organism) and then are either expelled out of the cell (as discussed previously for safety valves and multidrug resistance) or catabolised, with recovery of acetate or succinate at the end of the pathway (see, e.g. for the degradation of ornithine in *Pseudomonas aeruginosa* (Vander Wauven, Jann, Haas, Leisinger, & Stalon, 1988)).

Trash: Inactivation requires specific enzymes. In many cases, the most convenient versatile way of getting rid of a potentially toxic intermediate is to modify it to prevent it from entering core metabolic pathways and then to export the molecule out of the cell. This is most likely another reason, besides the control of osmotic pressure discussed previously, for the wealth of acetyltransferases, kinases, and methyltransferases (as modification enzymes) as well as 'multidrug transporters' that are encoded in most genomes. This preset functionality is at the root of antibiotic resistance and cancer drug resistance. Generally overlooked, it needs to be well understood and put under proper control in SB constructs.

Proofread: While this function is essential for macromolecules because of their genetic and energy cost, it is not frequently used in the case of metabolites except when chemical errors are frequently encountered. This is the case of errors in synthesis of proper stereoisomers (e.g. L-amino acids versus D-amino acids) (Van Schaftingen et al., 2013).

Repair: A similar view can be held for repair, and, interestingly, some metabolites are indeed repaired. However, only very costly metabolites, such as vitamin B12 or S-adenosylmethionine (AdoMet), are repaired rather than disposed of (Vinci & Clarke, 2007, Belda et al., 2013).

Salvage: The turnover of macromolecules and core metabolites results in compounds that are partially altered forms of the regular metabolite. In particular, purines and pyrimidines are salvaged in pathways that are important for nitrogen metabolism (Becerra & Lazcano, 1998; Hughes, Beck, & O'Donovan, 2005; Beck & O'Donovan, 2008). The salvaging of key molecules is further illustrated by a remarkable machinery, the degradosome in Bacteria and the exosome in Eukarya, that recycles RNA, in a way that keeps the concentration balance of nucleotides. Phosphorolysis is used as a way to spare energy while producing the essential precursor of cytidine in DNA, CDP. In parallel, a specific organisation of enzymes coupled to the degradosome, nucleoside diphosphokinase and pyruvate kinase, associated to enolase, replenishes the pool of GTP (Nitschke et al., 1998). The synthesis of the universal metabolite AdoMet is extremely costly, in particular because of the cost of the synthesis of methionine. Yet, AdoMet is used in the synthesis of essential polyamines, in a pathway that would result in the loss of an important amount of methionine, in the absence of a complex, but widespread, methionine salvage pathway (Sekowska et al., 2004).

Because SB most often deals with metabolic engineering, all these functions have to be explored with particular care, noticing, in particular, that many among them are almost ubiquitous.

4 PERSPECTIVE: THE FOURTH DIMENSION (WHEN TIME MEASURES AND SHAPES)

We reach the end of this chapter of the constraints that operate on the chasses used for SB applications. A final constraint, seldom considered, has yet to be explored. Adding *time* to the standard 3D structures is essential for SB engineering. We have already considered allostery and regulation of gene expression, where time is implicit but significant. Yet, there is a domain where the importance of time progressively becomes more prominent, that of maturation/ageing of biological structures.

Clocks, measuring time: Proteins are generally stable, but, as we saw previously, at least two of their residues, aspartate and asparagine, tend to cyclise (Figure 2.2). If this happens in proteins that are not submitted to fast degradation/resynthesis turnover, the modification acts as a protein-specific clock. The existence of multiple clocks is essential to smoothly run the genetic program since it works in a highly parallel fashion. Further types of clocks exist. Let us, for example, consider prion-related proteins such as those making structures called GW/P bodies, that are necessary for controlling RNA functions. Such proteins with prion-like domains assemble together to create temporary RNP granules (Moser & Fritzler, 2013). Upon meeting stress, RNP granules shut down the use of genes unrelated to stress by trapping the relevant RNAs. When conditions improve, the granules break apart. The glycine-rich flexible domain in the prion-like regions interact with RNA, but may also be prone to forming aggregates that develop in an autocatalytic way, creating another type of clock that, when unable to reset, triggers an irreversible damage to their host cells, possibly leading to cell death. This type of behaviour must be tightly controlled in SB constructs.

Time-dependent deformation: Besides acting as a clock, a time-dependent structure may change in time, reaching its final form at a proper location. While this concept is not yet explored in-depth in biology, this property is engineered with objects that have a 'shape-memory' based on pseudoelasticity of the material used

FIGURE 2.3

Self-shaping cube illustrating the function of a four-dimensions printer. The MIT's self-assembly laboratory and Stratasys Ltd. have designed an object that can be created by a 3D printer, and that subsequently changes shape over time, as illustrated in the three snapshots (http://www.sjet.us/MIT_4D%20PRINTING.html). As in proteins the initial tube can be coated with water, and it is made of a material that when placed at proper positions (which a 3D printer can easily program) will respond to water by bending spontaneously. In this figure a linear tube progressively folds as a cube. Proteins have similar properties, and it can certainly be accepted that structures extruded from a ribosome will be able to change shape after some pre-programmed time, in particular in specific local context, for example, resulting in local hardening of a functional shape.

(Luu, David, Ninomiya, & Winklbauer, 2011). A fourth-dimension printer creates a 3D structure that changes in time, to reach its final form, self-folding. It combines a strand of standard plastic with a layer made from a 'smart' material that can absorb water (Figure 2.3).

Exploration is the ultimate time-dependent master function of living organisms. This function allows them to colonise novel niches and to develop thriving populations. It also allows them to escape predators, flee toxic or dangerous conditions, and exchange genetic information with other organisms. Exploration is managed by a variety of specialised structures, for example, spores, pili, and fimbria. In most cases, its role is to cope with an unpredictable environment, that is, a situation that is far from the expected situation for engineering constructs. For this reason, the genetic setup that allows these structures to be implemented is most generally dispensable for SB constructs. It is likely to be good practice to delete the genes corresponding to exploration functions to prevent unwanted trapping and accumulation of contextual information by the designed cell factories.

SB must consider the chassis used to express human-designed programs. Top-down functional analysis is the way to trace many of the hidden functions that have been selected through the many generations of the evolution of life, by endeavouring to identify all that should not be omitted were we to (re)construct life. A main obstacle to be overcome is the in-built 'inventive' activity of living systems that tend to extract information from the environment on a systematic basis. This feature is backed by energy-dependent degradation or sorting systems that have been shaped to allow propagation of the organism in an unpredictable future. A major aim of the future of large SB setups will be to prepare these systems to fit with human goals.

Acknowledgements

We wish to thank Eric Fourmentin for constructive critical comments on this manuscript. This work benefited from ongoing discussions of the Stanislas Noria Network. It has been supported by the FP7 European Union programme Microme KBBE-2007-3-2-08-222886-2 grant.

References

Acevedo-Rocha, C. G., Fang, G., Schmidt, M., Ussery, D. W., & Danchin, A. (2013). From essential to persistent genes: A functional approach to constructing synthetic life. *Trends in Genetics, 29,* 273–279.

Achbergerova, L., & Nahalka, J. (2011). Polyphosphate—An ancient energy source and active metabolic regulator. *Microbial Cell Factories, 10,* 63.

Ackermann, H. W. (1998). Tailed bacteriophages: The order caudovirales. *Advances in Virus Research, 51,* 135–201.

Agapakis, C. M., Boyle, P. M., & Silver, P. A. (2012). Natural strategies for the spatial optimization of metabolism in synthetic biology. *Nature Chemical Biology, 8,* 527–535.

Ang, J., & McMillen, D. R. (2013). Physical constraints on biological integral control design for homeostasis and sensory adaptation. *Biophysical Journal, 104,* 505–515.

Aravind, L. (2000). Guilt by association: Contextual information in genome analysis. *Genome Research, 10,* 1074–1077.

Bachman, B. J., Venner, E., Lua, R. C., Erdin, S., & Lichtarge, O. (2012). ETAscape: Analyzing protein networks to predict enzymatic function and substrates in Cytoscape. *Bioinformatics, 28,* 2186–2188.

Bailly-Bechet, M., Danchin, A., Iqbal, M., Marsili, M., & Vergassola, M. (2006). Codon usage domains over bacterial chromosomes. *PLoS Computational Biology, 2,* e37.

Balakrishnan, L., & Bambara, R. A. (2013). Okazaki fragment metabolism. *Cold Spring Harbor Perspectives in Biology, 5,* a010173.

Barth, P., Schoeffler, A., & Alber, T. (2008). Targeting metastable coiled-coil domains by computational design. *Journal of the American Chemical Society, 130,* 12038–12044.

Becerra, A., & Lazcano, A. (1998). The role of gene duplication in the evolution of purine nucleotide salvage pathways. *Origins of Life and Evolution of the Biosphere, 28,* 539–553.

Beck, D. A., & O'Donovan, G. A. (2008). Pathways of pyrimidine salvage in Pseudomonas and former Pseudomonas: Detection of recycling enzymes using high-performance liquid chromatography. *Current Microbiology, 56,* 162–167.

Belda, E., Sekowska, A., Le Fevre, F., Morgat, A., Mornico, D., Ouzounis, C., et al. (2013). An updated metabolic view of the *Bacillus subtilis* 168 genome. *Microbiology, 159,* 757–770.

Bergey's Manual Trust, (1986). *Bergey's manual of systematic bacteriology.* Baltimore: Williams & Wilkins Co.

Berthoumieux, S., de Jong, H., Baptist, G., Pinel, C., Ranquet, C., Ropers, D., et al. (2013). Shared control of gene expression in bacteria by transcription factors and global physiology of the cell. *Molecular Systems Biology, 9,* 634.

Binder, P. M., & Danchin, A. (2011). Life's demons: Information and order in biology. What subcellular machines gather and process the information necessary to sustain life? *EMBO Reports, 12,* 495–499.

Bogumil, D., & Dagan, T. (2012). Cumulative impact of chaperone-mediated folding on genome evolution. *Biochemistry, 51*, 9941–9953.

Bonacci, W., Teng, P. K., Afonso, B., Niederholtmeyer, H., Grob, P., Silver, P. A., et al. (2012). Modularity of a carbon-fixing protein organelle. *Proceedings of the National Academy of Sciences of the United States of America, 109*, 478–483.

Brenner, S. (2012). Turing centenary: Life's code script. *Nature, 482*, 461.

Callewaert, L., Van Herreweghe, J. M., Vanderkelen, L., Leysen, S., Voet, A., & Michiels, C. W. (2012). Guards of the great wall: Bacterial lysozyme inhibitors. *Trends in Microbiology, 20*, 501–510.

Cascales, E., Buchanan, S. K., Duche, D., Kleanthous, C., Lloubes, R., Postle, K., et al. (2007). Colicin biology. *Microbiology and Molecular Biology Reviews, 71*, 158–229.

Celler, K., Koning, R. I., Koster, A. J., & van Wezel, G. P. (2013). A multi-dimensional view of the bacterial cytoskeleton. *Journal of Bacteriology, 195*, 1627–1636.

Chandra, G., Chater, K. F., & Bornemann, S. (2011). Unexpected and widespread connections between bacterial glycogen and trehalose metabolism. *Microbiology, 157*, 1565–1572.

Chlebowski, A., Lubas, M., Jensen, T. H., & Dziembowski, A. (2013). RNA decay machines: The exosome. *Biochimica et Biophysica Acta, 1829*, 552–560.

Chong, P. L., Ayesa, U., Daswani, V. P., & Hur, E. C. (2012). On physical properties of tetraether lipid membranes: Effects of cyclopentane rings. *Archaea, 2012*, 138439.

Chuang, J. S. (2012). Engineering multicellular traits in synthetic microbial populations. *Current Opinion in Chemical Biology, 16*, 370–378.

Cloutier, P., & Coulombe, B. (2013). Regulation of molecular chaperones through post-translational modifications: Decrypting the chaperone code. *Biochimica et Biophysica Acta, 1829*, 443–454.

Cole, E. L. Jr. (1998). Functional analysis: A system conceptual design tool [and application to ATC system]. *IEEE Transactions on Aerospace and Electronic Systems, 34*, 354–365.

Copley, S. J. Giannarou, S., Schmid, V. J., Hansell, D. M., Wells, A. U., & Yang, G. Z. (2012). Effect of aging on lung structure in vivo: Assessment with densitometric and fractal analysis of high-resolution computed tomography data. *Journal of Thoracic Imaging, 27*, 366–371.

Corry, B., & Thomas, M. (2012). Mechanism of ion permeation and selectivity in a voltage gated sodium channel. *Journal of the American Chemical Society, 134*, 1840–1846.

da Silva, K. P., & Borges, J. C. (2010). The molecular chaperone Hsp70 family members function by a bidirectional heterotrophic allosteric mechanism. *Protein & Peptide Letters, 18*, 132–142.

Danchin, A. (1999). From protein sequence to function. *Current Opinion in Structural Biology, 9*, 363–367.

Danchin, A. (2003). *The Delphic boat. What genomes tell us.* (Quayle, A., Trans.). Cambridge, MA: Harvard University Press.

Danchin, A. (2009a). A phylogenetic view of bacterial ribonucleases. *Progress in Molecular Biology and Translational Science, 85*, 1–41.

Danchin, A. (2009b). Bacteria as computers making computers. *FEMS Microbiology Reviews, 33*, 3–26.

Danchin, A. (2009c). Cells need safety valves. *Bioessays, 31*, 769–773.

Danchin, A. (2009d). Myopic selection of novel information drives evolution. *Current Opinion in Biotechnology, 20*, 504–508.

Danchin, A. (2012). Scaling up synthetic biology: Do not forget the chassis. *FEBS Letters, 586*, 2129–2137.

Danchin, A., Binder, P. M., & Noria, S. (2011). Antifragility and tinkering in biology (and in business) flexibility provides an efficient epigenetic way to manage risk. *Genes, 2,* 998–1016.

Danchin, A., & Dondon, L. (1979). Regulatory features of tRNA Leu I expression in *Escherichia coli* K12. *Biochemical and Biophysical Research Communications, 90,* 1280–1286.

Danchin, A., Guerdoux-Jamet, P., Moszer, I., & Nitschké, P. (2000). Mapping the bacterial cell architecture into the chromosome. *Philosophical Transactions of the Royal Society of London. Series B, Biological Sciences, 355,* 179–190.

Danielli, J. F. (1972). Context and future of cell synthesis. *New York State Journal of Medicine, 72,* 2814–2815.

Danielli, J. F. (1974). Genetic engineering and life synthesis: An introduction to the review by R. Widdus and C. Ault. *International Review of Cytology, 38,* 1–5.

David, C. L., Keener, J., & Aswad, D. W. (1999). Isoaspartate in ribosomal protein S11 of Escherichia coli. *Journal of Bacteriology, 181,* 2872–2877.

Davies, J., & Davies, D. (2010). Origins and evolution of antibiotic resistance. *Microbiology and Molecular Biology Reviews, 74,* 417–433.

de Crecy-Lagard, V. A., Bellalou, J., Mutzel, R., & Marliere, P. (2001). Long term adaptation of a microbial population to a permanent metabolic constraint: Overcoming thymineless death by experimental evolution of *Escherichia coli*. *BMC Biotechnology, 1,* 10.

Desloover, J., Arends, J. B., Hennebel, T., & Rabaey, K. (2012). Operational and technical considerations for microbial electrosynthesis. *Biochemical Society Transactions, 40,* 1233–1238.

Dunker, A. K., Garner, E., Guilliot, S., Romero, P., Albrecht, K., Hart, J., et al. (1998). Protein disorder and the evolution of molecular recognition: Theory, predictions and observations. *Pacific Symposium on Biocomputing,* 473–484.

Dykhuizen, D., & Hartl, D. (1978). Transport by the lactose permease of *Escherichia coli* as the basis of lactose killing. *Journal of Bacteriology, 135,* 876–882.

Ellis, J. (1987). Proteins as molecular chaperones. *Nature, 328,* 378–379.

Ellis, T., Adie, T., & Baldwin, G. S. (2011). DNA assembly for synthetic biology: From parts to pathways and beyond. *Integrative Biology, 3,* 109–118.

Elowitz, M. B., & Leibler, S. (2000). A synthetic oscillatory network of transcriptional regulators. *Nature, 403,* 335–338.

Endy, D. (2005). Foundations for engineering biology. *Nature, 438,* 449–453.

Engelen, S., Vallenet, D., Medigue, C., & Danchin, A. (2011). Distinct co-evolution patterns of genes associated to DNA polymerase III DnaE and PolC. *BMC Genomics, 13,* 69.

Fang, G., Rocha, E., & Danchin, A. (2005). How essential are nonessential genes? *Molecular Biology and Evolution, 22,* 2147–2156.

Fantoni, G., Apreda, R., & Bonaccorsi, A. (2009). A theory of the constituent elements of functions. In *Paper presented at the proceedings of the 17th international conference on engineering design (ICED'09), Stanford University, San Francisco, CA, USA*.

Fijalkowska, I. J., Schaaper, R. M., & Jonczyk, P. (2012). DNA replication fidelity in *Escherichia coli*: A multi-DNA polymerase affair. *FEMS Microbiology Reviews, 36,* 1105–1121.

Forster, A. C., & Church, G. M. (2006). Towards synthesis of a minimal cell. *Molecular Systems Biology, 2,* 45.

Freist, W., Sternbach, H., Pardowitz, I., & Cramer, F. (1998). Accuracy of protein biosynthesis: Quasi-species nature of proteins and possibility of error catastrophes. *Journal of Theoretical Biology, 193,* 19–38.

Fu, Q., Kobayashi, H., Kawaguchi, H., Vilcaez, J., Wakayama, T., Maeda, H., et al. (2013). Electrochemical and phylogenetic analyses of current-generating microorganisms in a thermophilic microbial fuel cell. *Journal of Bioscience and Bioengineering*, *115*, 268–271.

Fukui, K. (2010). DNA mismatch repair in eukaryotes and bacteria. *Journal of Nucleic Acids*, *2010*, ID260512.

Galperin, M. Y., Walker, D. R., & Koonin, E. V. (1998). Analogous enzymes: Independent inventions in enzyme evolution. *Genome Research*, *8*, 779–790.

Gardner, T. S., Cantor, C. R., & Collins, J. J. (2000). Construction of a genetic toggle switch in Escherichia coli. *Nature*, *403*, 339–342.

Gendrault, Y., Madec, M., Wlotzko, V., Andraud, M., Lallement, C., & Haiech, J. (2012). Using digital electronic design flow to create a Genetic Design Automation tool. *Conference Proceedings: Annual International Conference of the IEEE Engineering in Medicine and Biology Society*, *2012*, 5530–5533.

Gibson, L. J. (2012). The hierarchical structure and mechanics of plant materials. *Journal of the Royal Society, Interface*, *9*, 2749–2766.

Gillis, J., & Pavlidis, P. (2011). The impact of multifunctional genes on "guilt by association" analysis. *PLoS One*, *6*, e17258.

Gillis, J., & Pavlidis, P. (2012). "Guilt by association" is the exception rather than the rule in gene networks. *PLoS Computational Biology*, *8*, e1002444.

Giraut, A., Abu El-Asrar, R., Marliere, P., Delarue, M., & Herdewijn, P. (2012). 2′-Deoxyribonucleoside phosphoramidate triphosphate analogues as alternative substrates for *E. coli* polymerase III. *Chembiochem*, *13*, 2439–2444.

Greene, S. E., & Komeili, A. (2012). Biogenesis and subcellular organization of the magnetosome organelles of magnetotactic bacteria. *Current Opinion in Cell Biology*, *24*, 490–495.

Grohman, J. K., Gorelick, R. J., Lickwar, C. R., Lieb, J. D., Bower, B. D., Znosko, B. M., et al. (2013). A guanosine-centric mechanism for RNA chaperone function. *Science*, *340*, 190–195.

Guerrero, R., & Berlanga, M. (2007). The hidden side of the prokaryotic cell: Rediscovering the microbial world. *International Microbiology*, *10*, 157–168.

Gupta, R. S. (2011). Origin of diderm (Gram-negative) bacteria: Antibiotic selection pressure rather than endosymbiosis likely led to the evolution of bacterial cells with two membranes. *Antonie Van Leeuwenhoek*, *100*, 171–182.

Hanson, B. R., & Neely, M. N. (2012). Coordinate regulation of Gram-positive cell surface components. *Current Opinion in Microbiology*, *15*, 204–210.

Harold, F. M. (2007). Bacterial morphogenesis: Learning how cells make cells. *Current Opinion in Microbiology*, *10*, 591–595.

Hartfield, M., & Keightley, P. D. (2012). Current hypotheses for the evolution of sex and recombination. *Integrative Zoology*, *7*, 192–209.

Haswell, E. S., Phillips, R., & Rees, D. C. (2011). Mechanosensitive channels: What can they do and how do they do it? *Structure*, *19*, 1356–1369.

Hegler, J. A., Weinkam, P., & Wolynes, P. G. (2008). The spectrum of biomolecular states and motions. *Human Frontier Science Program*, *2*, 307–313.

Heizer, E. M., Jr., Raiford, D. W., Raymer, M. L., Doom, T. E., Miller, R. V., & Krane, D. E. (2006). Amino acid cost and codon-usage biases in 6 prokaryotic genomes: A whole-genome analysis. *Molecular Biology and Evolution*, *23*, 1670–1680.

Heukeshoven, J., Marz, A., Warnecke, G., Deppert, W., & Tolstonog, G. V. (2012). Recombinant p53 displays heterogeneity during isoelectric focusing. *Electrophoresis*, *33*, 2818–2827.

Hoffmann, A., Bukau, B., & Kramer, G. (2010). Structure and function of the molecular chaperone Trigger Factor. *Biochimica et Biophysica Acta, 1803*, 650–661.

Hopfield, J. J. (1974). Kinetic proofreading: A new mechanism for reducing errors in biosynthetic processes requiring high specificity. *Proceedings of the National Academy of Sciences of the United States of America, 71*, 4135–4139.

Hopkins, J. D. (1974). A new class of promoter mutations in the lactose operon of Escherichia coli. *Journal of Molecular Biology, 87*, 715–724.

Hughes, L. E., Beck, D. A., & O'Donovan, G. A. (2005). Pathways of pyrimidine salvage in Streptomyces. *Current Microbiology, 50*, 8–10.

Huynh, T. N., & Stewart, V. (2011). Negative control in two-component signal transduction by transmitter phosphatase activity. *Molecular Microbiology, 82*, 275–286.

Jacob, F. (1974). *The logic of life: A history of heredity.* New York: Pantheon Books.

Johnson, K. A. (2008). Role of induced fit in enzyme specificity: A molecular forward/reverse switch. *The Journal of Biological Chemistry, 283*, 26297–26301.

Juhas, M., Eberl, L., & Glass, J. I. (2011). Essence of life: Essential genes of minimal genomes. *Trends in Cell Biology, 21*, 562–568.

Jun, S., & Mulder, B. (2006). Entropy-driven spatial organization of highly confined polymers: Lessons for the bacterial chromosome. *Proceedings of the National Academy of Sciences of the United States of America, 103*, 12388–12393.

Jung, K., Fried, L., Behr, S., & Heermann, R. (2012). Histidine kinases and response regulators in networks. *Current Opinion in Microbiology, 15*, 118–124.

Kenney, L. J. (2010). How important is the phosphatase activity of sensor kinases? *Current Opinion in Microbiology, 13*, 168–176.

Khudyakov, I. Y., Kirnos, M. D., Alexandrushkina, N. I., & Vanyushin, B. F. (1978). Cyanophage S-2L contains DNA with 2,6-diaminopurine substituted for adenine. *Virology, 88*, 8–18.

Khunjar, W. O., Sahin, A., West, A. C., Chandran, K., & Banta, S. (2012). Biomass production from electricity using ammonia as an electron carrier in a reverse microbial fuel cell. *PLoS One, 7*, e44846.

Kroh, L. W., Fiedler, T., & Wagner, J. (2008). alpha-Dicarbonyl compounds—Key intermediates for the formation of carbohydrate-based melanoidins. *Annals of the New York Academy of Sciences, 1126*, 210–215.

Lagesen, K., Ussery, D. W., & Wassenaar, T. M. (2010). Genome update: The 1000th genome—A cautionary tale. *Microbiology, 156*, 603–608.

Lambrecht, J. A., Schmitz, G. E., & Downs, D. M. (2013). RidA proteins prevent metabolic damage inflicted by PLP-dependent dehydratases in all domains of life. *MBio, 4*, e00033.

Laskey, R. A., Honda, B. M., Mills, A. D., & Finch, J. T. (1978). Nucleosomes are assembled by an acidic protein which binds histones and transfers them to DNA. *Nature, 275*, 416–420.

Lehnik-Habrink, M., Pfortner, H., Rempeters, L., Pietack, N., Herzberg, C., & Stulke, J. (2010). The RNA degradosome in *Bacillus subtilis*: Identification of CshA as the major RNA helicase in the multiprotein complex. *Molecular Microbiology, 77*, 958–971.

Lifson, S., & Zimm, B. H. (1963). Simplified theory of the helix–coil transition in DNA based on a grand partition function. *Biopolymers, 1*, 15–23.

Liu, M. F., Cescau, S., Mechold, U., Wang, J., Cohen, D., Danchin, A., et al. (2012). Identification of a novel nanoRNase in Bartonella. *Microbiology, 158*, 886–895.

Lloyd-Price, J., Hakkinen, A., Kandhavelu, M., Marques, I. J., Chowdhury, S., Lihavainen, E., et al. (2012). Asymmetric disposal of individual protein aggregates in *Escherichia coli*, one aggregate at a time. *Journal of Bacteriology, 194*, 1747–1752.

Luu, O., David, R., Ninomiya, H., & Winklbauer, R. (2011). Large-scale mechanical properties of Xenopus embryonic epithelium. *Proceedings of the National Academy of Sciences of the United States of America*, *108*, 4000–4005.

Maisonneuve, E., Fraysse, L., Moinier, D., & Dukan, S. (2008). Existence of abnormal protein aggregates in healthy E*scherichia coli* cells. *Journal of Bacteriology*, *190*, 887–893.

Mammoto, A., Mammoto, T., & Ingber, D. E. (2012). Mechanosensitive mechanisms in transcriptional regulation. *Journal of Cell Science*, *125*, 3061–3073.

Marinus, M. G., & Casadesus, J. (2009). Roles of DNA adenine methylation in host–pathogen interactions: Mismatch repair, transcriptional regulation, and more. *FEMS Microbiology Reviews*, *33*, 488–503.

Maxwell, J. C. (1871). *Theory of heat*. London: Longmans, Reed and Co.

Mingorance, J., Tamames, J., & Vicente, M. (2004). Genomic channeling in bacterial cell division. *Journal of Molecular Recognition*, *17*, 481–487.

Moser, J. J., & Fritzler, M. J. (2013). Relationship of other cytoplasmic ribonucleoprotein bodies (cRNPB) to GW/P bodies. *Advances in Experimental Medicine and Biology*, *768*, 213–242.

Motlagh, H. N., Li, J., Thompson, E. B., & Hilser, V. J. (2012). Interplay between allostery and intrinsic disorder in an ensemble. *Biochemical Society Transactions*, *40*, 975–980.

Nitschke, P., Guerdoux-Jamet, P., Chiapello, H., Faroux, G., Henaut, C., Henaut, A., et al. (1998). Indigo: A World-Wide-Web review of genomes and gene functions. *FEMS Microbiology Reviews*, *22*, 207–227.

Noria, S., & Danchin, A. (2002). Just so genome stories: What does my neighbor tell me? In *Paper presented at the Uehara Memorial Foundation Symposium: Genome Science: Towards a new paradigm? Tokyo*.

Norton, J. P., & Mulvey, M. A. (2012). Toxin-antitoxin systems are important for niche-specific colonization and stress resistance of uropathogenic Escherichia coli. *PLoS Pathogens*, *8*, e1002954.

Nussinov, R., Tsai, C. J., & Ma, B. (2013). The (still) underappreciated role of allostery in the cellular network. *Annual Review of Biophysics*, *42*, 169–189.

Overbeek, R., Fonstein, M., D'Souza, M., Pusch, G. D., & Maltsev, N. (1999). Use of contiguity on the chromosome to predict functional coupling. *In Silico Biology*, *1*, 93–108.

Page, M. G. (2012). The role of the outer membrane of Gram-negative bacteria in antibiotic resistance: Ajax' shield or Achilles' heel? *Handbook of Experimental Pharmacology*, *211*, 67–86.

Pascal, G., Medigue, C., & Danchin, A. (2005). Universal biases in protein composition of model prokaryotes. *Proteins*, *60*, 27–35.

Pavkov-Keller, T., Howorka, S., & Keller, W. (2011). The structure of bacterial S-layer proteins. *Progress in Molecular Biology and Translational Science*, *103*, 73–130.

Pazos, F., & Valencia, A. (2008). Protein co-evolution, co-adaptation and interactions. *EMBO Journal*, *27*, 2648–2655.

Petersen, S. B., Neves-Petersen, M. T., Henriksen, S. B., Mortensen, R. J., & Geertz-Hansen, H. M. (2012). Scale-free behaviour of amino acid pair interactions in folded proteins. *PLoS One*, *7*, e41322.

Pfeffer, C., Larsen, S., Song, J., Dong, M., Besenbacher, F., Meyer, R. L., et al. (2012). Filamentous bacteria transport electrons over centimetre distances. *Nature*, *491*, 218–221.

Pilhofer, M., & Jensen, G. J. (2013). The bacterial cytoskeleton: More than twisted filaments. *Current Opinion in Cell Biology*, *25*, 125–133.

Pommier, Y. (2013). Drugging topoisomerases: Lessons and challenges. *ACS Chemical Biology*, *8*, 82–95.

Preissler, S., & Deuerling, E. (2012). Ribosome-associated chaperones as key players in proteostasis. *Trends in Biochemical Sciences*, *37*, 274–283.

Przulj, N., Corneil, D. G., & Jurisica, I. (2004). Modeling interactome: Scale-free or geometric? *Bioinformatics*, *20*, 3508–3515.

Qi, L. S., Larson, M. H., Gilbert, L. A., Doudna, J. A., Weissman, J. S., Arkin, A. P., et al. (2013). Repurposing CRISPR as an RNA-guided platform for sequence-specific control of gene expression. *Cell*, *152*, 1173–1183.

Rauhut, R., & Klug, G. (1999). mRNA degradation in bacteria. *FEMS Microbiology Reviews*, *23*, 353–370.

Redko, Y., Aubert, S., Stachowicz, A., Lenormand, P., Namane, A., Darfeuille, F., et al. (2013). A minimal bacterial RNase J-based degradosome is associated with translating ribosomes. *Nucleic Acids Research*, *41*, 288–301.

Reese, T. M., Brzoska, A., Yott, D. T., & Kelleher, D. J. (2012). Analyzing self-similar and fractal properties of the C. elegans neural network. *PLoS One*, *7*, e40483.

Riccione, K. A., Smith, R. P., Lee, A. J., & You, L. (2013). A synthetic biology approach to understanding cellular information processing. *ACS Synthetic Biology*, *1*, 389–402.

Robinson, N., & Robinson, A. (2004). *Molecular Clocks Deamidation of asparaginyl and glutaminyl residues in peptides and proteins*. Cave Junction, Oregon: Althouse Press.

Rocha, E. P., & Danchin, A. (2003). Gene essentiality determines chromosome organisation in bacteria. *Nucleic Acids Research*, *31*, 6570–6577.

Rocha, E. P., Guerdoux-Jamet, P., Moszer, I., Viari, A., & Danchin, A. (2000). Implication of gene distribution in the bacterial chromosome for the bacterial cell factory. *Journal of Biotechnology*, *78*, 209–219.

Rodrigo-Brenni, M. C., & Hegde, R. S. (2012). Design principles of protein biosynthesis-coupled quality control. *Developmental Cell*, *23*, 896–907.

Romero, P., Obradovic, Z., Kissinger, C. R., Villafranca, J. E., Garner, E., Guilliot, S., et al. (1998). Thousands of proteins likely to have long disordered regions. *Pacific Symposium on Biocomputing*, 437–448.

Saberi, S., & Emberly, E. (2010). Chromosome driven spatial patterning of proteins in bacteria. *PLoS Computational Biology*, *6*, e1000986.

Schiene-Fischer, C., Aumuller, T., & Fischer, G. (2013). Peptide bond cis/trans isomerases: A biocatalysis perspective of conformational dynamics in proteins. *Topics in Current Chemistry*, *328*, 35–67.

Sekowska, A., Denervaud, V., Ashida, H., Michoud, K., Haas, D., Yokota, A., et al. (2004). Bacterial variations on the methionine salvage pathway. *BMC Microbiology*, *4*, 9.

Seufferheld, M. J., Kim, K. M., Whitfield, J., Valerio, A., & Caetano-Anolles, G. (2011). Evolution of vacuolar proton pyrophosphatase domains and volutin granules: Clues into the early evolutionary origin of the acidocalcisome. *Biology Direct*, *6*, 50.

Shen-Orr, S. S., Milo, R., Mangan, S., & Alon, U. (2002). Network motifs in the transcriptional regulation network of *Escherichia coli*. *Nature Genetics*, *31*, 64–68.

Shetty, R. P., Endy, D., & Knight, T. F. Jr. (2008). Engineering BioBrick vectors from BioBrick parts. *Journal of Biological Engineering*, *2*, 5.

Shimura, K., Hoshino, M., Kamiya, K., Enomoto, M., Hisada, S., Matsumoto, H., et al. (2013). Estimation of the deamidation rates of major deamidation sites in a Fab fragment of mouse IgG1-kappa by capillary isoelectric focusing of mutated Fab fragments. *Analytical Chemistry*, *85*, 1705–1710.

Silva-Rocha, R., Martinez-Garcia, E., Calles, B., Chavarria, M., Arce-Rodriguez, A., de Las Heras, A., et al. (2013). The Standard European Vector Architecture (SEVA): A coherent platform for the analysis and deployment of complex prokaryotic phenotypes. *Nucleic Acids Research, 41*, D666–D675.

Singh, S. P., & Montgomery, B. L. (2011). Determining cell shape: Adaptive regulation of cyanobacterial cellular differentiation and morphology. *Trends in Microbiology, 19*, 278–285.

Smith, G. R. (2012). How RecBCD enzyme and Chi promote DNA break repair and recombination: A molecular biologist's view. *Microbiology and Molecular Biology Reviews, 76*, 217–228.

Souza, W. (2012). Prokaryotic cells: Structural organisation of the cytoskeleton and organelles. *Memórias do Instituto Oswaldo Cruz, 107*, 283–293.

Sporns, O. (2006). Small-world connectivity, motif composition, and complexity of fractal neuronal connections. *Biosystems, 85*, 55–64.

Stewart, E. J., Madden, R., Paul, G., & Taddei, F. (2005). Aging and death in an organism that reproduces by morphologically symmetric division. *PLoS Biology, 3*, e45.

Stoop, E. J., Bitter, W., & van der Sar, A. M. (2012). Tubercle bacilli rely on a type VII army for pathogenicity. *Trends in Microbiology, 20*, 477–484.

Tamames, J., Gonzalez-Moreno, M., Mingorance, J., Valencia, A., & Vicente, M. (2001). Bringing gene order into bacterial shape. *Trends in Genetics, 17*, 124–126.

Tamaru, Y., Miyake, H., Kuroda, K., Nakanishi, A., Matsushima, C., Doi, R. H., et al. (2011). Comparison of the mesophilic cellulosome-producing *Clostridium cellulovorans* genome with other cellulosome-related clostridial genomes. *Microbial Biotechnology, 4*, 64–73.

Thom, R. (1989). *Structural stability and morphogenesis: An outline of a general theory of models (translated from French: Stabilité structurelle et morphogenèse, 1072) (Fowler, D. H., Trans.).* Reading, MA: Addison-Wesley Publishers.

Thomas, M., Jayatilaka, D., & Corry, B. (2013). An entropic mechanism of generating selective ion binding in macromolecules. *PLoS Computational Biology, 9*, e1002914.

Thompson, R. C. (1988). EFTu provides an internal kinetic standard for translational accuracy. *Trends in Biochemical Sciences, 13*, 91–93.

Tilly, K., & Georgopoulos, C. (1982). Evidence that the two *Escherichia coli groE* morphogenetic gene products interact in vivo. *Journal of Bacteriology, 149*, 1082–1088.

Timsit, Y., Allemand, F., Chiaruttini, C., & Springer, M. (2006). Coexistence of two protein folding states in the crystal structure of ribosomal protein L20. *EMBO Reports, 7*, 1013–1018.

Tzeng, S. R., & Kalodimos, C. G. (2011). Protein dynamics and allostery: An NMR view. *Current Opinion in Structural Biology, 21*, 62–67.

Van Schaftingen, E., Rzem, R., Marbaix, A., Collard, F., Veiga-da-Cunha, M., & Linster, C. L. (2013). Metabolite proofreading, a neglected aspect of intermediary metabolism. *Journal of Inherited Metabolic Disease, 36*, 427–434.

Vander Wauven, C., Jann, A., Haas, D., Leisinger, T., & Stalon, V. (1988). N2-succinylornithine in ornithine catabolism of *Pseudomonas aeruginosa*. *Archives of Microbiology, 150*, 400–404.

Varela, M. F., Brooker, R. J., & Wilson, T. H. (1997). Lactose carrier mutants of *Escherichia coli* with changes in sugar recognition (lactose versus melibiose). *Journal of Bacteriology, 179*, 5570–5573.

Vinci, C. R., & Clarke, S. G. (2007). Recognition of age-damaged (R, S)-adenosyl-L-methionine by two methyltransferases in the yeast *Saccharomyces cerevisiae*. *Journal of Biological Chemistry*, *282*, 8604–8612.

Wang, B., & Buck, M. (2012). Customizing cell signaling using engineered genetic logic circuits. *Trends in Microbiology*, *20*, 376–384.

Wang, P., Robert, L., Pelletier, J., Dang, W. L., Taddei, F., Wright, A., et al. (2010). Robust growth of *Escherichia coli*. *Current Biology*, *20*, 1099–1103.

Wang, S., & Shaevitz, J. W. (2013). The mechanics of shape in prokaryotes. *Frontiers in Bioscience (Scholar Edition)*, *5*, 564–574.

Wang, S., & Wingreen, N. S. (2013). Cell shape can mediate the spatial organization of the bacterial cytoskeleton. *Biophysical Journal*, *104*, 541–552.

Willenbrock, H., & Ussery, D. W. (2004). Chromatin architecture and gene expression in Escherichia coli. *Genome Biology*, *5*, 252.

Wirth, S. E., Krywy, J. A., Aldridge, B. B., Fortune, S. M., Fernandez-Suarez, M., Gray, T. A., et al. (2012). Polar assembly and scaffolding proteins of the virulence-associated ESX-1 secretory apparatus in mycobacteria. *Molecular Microbiology*, *83*, 654–664.

Wittmann, A., & Suess, B. (2012). Engineered riboswitches. Expanding researchers' toolbox with synthetic RNA regulators. *FEBS Letters*, *586*, 2076–2083.

Yadavalli, S. S., & Ibba, M. (2012). Quality control in aminoacyl-tRNA synthesis its role in translational fidelity. *Advances in Protein Chemistry and Structural Biology*, *86*, 1–43.

Zhang, L. Y., Chang, S. H., & Wang, J. (2011). How to make a minimal genome for synthetic minimal cell. *Protein & Cell*, *1*, 427–434.

Zhao, L., Hoi, S. C., Wong, L., Hamp, T., & Li, J. (2012). Structural and functional analysis of multi-interface domains. *PLoS One*, *7*, e50821.

Zucca, S., Pasotti, L., Mazzini, G., De Angelis, M. G., & Magni, P. (2012). Characterization of an inducible promoter in different DNA copy number conditions. *BMC Bioinformatics*, *13* (Suppl. 4), S11.

CHAPTER 3

Social Dimensions of Microbial Synthetic Biology

Jane Calvert[1], Emma Frow

Science, Technology and Innovation Studies, University of Edinburgh, Old Surgeons' Hall, High School Yards, Edinburgh, United Kingdom
[1]*Corresponding author. e-mail address: Jane.Calvert@ed.ac.uk*

1 INTRODUCTION: MAKING SPACE FOR A NEW DISCUSSION

It is increasingly commonplace to find that conferences, reports and books on new technologies will devote a slot to the discussion of the 'ethical and social issues'. This is a fairly recent trend, one that began in the 1990s when the public funders of the Human Genome Project decided that 5% of the research budget should be devoted to studying the 'ethical, legal and social implications' (ELSI) of sequencing the human genome. The ELSI label has become institutionalised over the past two decades, with the result that humanities and social science researchers are increasingly being brought in to study and/or to collaborate with scientists and engineers on new and emerging areas of science and technology.

Synthetic biology is no exception. It is still a relatively young and small field (foundational papers include Gardner, Cantor, & Collins, 2000 and Elowitz & Leibler, 2000), but there have been a surprisingly large number of reports on the ethical, legal and social implications of synthetic biology—Zhang, Marris, and Rose (2011) identified 39 English-language reports from 2004 to 2011. These reports raise a fairly consistent set of 'issues' for discussion in relation to synthetic biology, typically couched under the headings of biosafety, biosecurity, intellectual property, 'creating life', and public acceptance and/or dialogue (see, e.g. ETC Group, 2007; Royal Academy of Engineering, 2009).

Rather than rehearsing these now well-documented issues, we have decided to take a slightly different approach with this chapter. We want to take a step back from some of the specific issues listed earlier and instead consider, in broader terms, the framing and scope of the 'social issues' around emerging technologies such as synthetic biology. We do this for two reasons. One is to provide some background and context as to how we think about the relationship between science and society—a perspective that can sometimes get buried when diving straight into a specific issue like the biosecurity threats posed by synthetic biology. The perspective we adopt is one of researchers grounded in the discipline of science and technology studies (or STS). STS researchers explore the social, political, economic, philosophical and historical dimensions of science and technology. This type of research shows us that science and society are not two separate spheres that influence one another from afar but that they are inseparable: science is part of society and society is part of science. Few people would deny the former, but the idea that society is part of science is perhaps less intuitive. We hope to illustrate this throughout the chapter by demonstrating the social, political and economic influences on the nature of scientific and technological work that is done and on the type of knowledge that is brought into the world.

The second reason for taking this approach to a 'social dimensions' chapter is because we sometimes find that the scientists, engineers and policymakers we work with want answers to questions that we feel might be somewhat problematic questions to begin with—for example, questions around how to ensure better public acceptance of this new and ground-breaking technology. In this chapter, we hope to show how and why the work of STS reveals this to be a problematic question, and we point to what could be potentially more productive avenues for discussion.

Thus, what we hope to do with this chapter is to challenge some of the more familiar ways of thinking and to propose a series of shifts in the framing and language used for discussing the social aspects of synthetic biology—moving away from discussions centred around ethical implications, speculations about the future and concerns about risk, regulation and public acceptance towards a conversation that talks in terms of social dimensions, anticipating the future, managing uncertainty, tools of governance and research for the public good. We argue that these seemingly subtle changes in vocabulary open up a new and productive space for thinking about the social dimensions of synthetic biology.

2 FROM IMPLICATIONS TO DIMENSIONS

Our first shift, reflected in the title of this chapter, is from using the word 'implications' to thinking instead about 'dimensions'. The word 'implications', that is, the 'I' of the ELSI label, can arguably encourage a problematic way of thinking—an assumption that social and ethical issues should be considered *after* the work of science and engineering has been completed (Fisher, 2005). Using this word makes it easy to think of scientific research as something separated or bracketed off from society. The study of social 'dimensions', in contrast, does not imply that this discussion needs to happen at the downstream end of research and development. Social dimensions can be identified and discussed throughout the research process—before, during and after laboratory work takes place. It is a more flexible and encompassing phrase that does not reinforce a science–society divide.

Because STS researchers are interested in dimensions rather than implications, they often work alongside or in collaboration with scientists and engineers. We have engaged with synthetic biologists in a variety of research, teaching and policy initiatives over the last 4 years. Rather than commenting on the prospects and practices of synthetic biology from a distance, we aim to gain a detailed understanding of, and be closely involved with, the issues that scientists and engineers are grappling with in real time.

2.1 Metaphors and analogies

To give an example of how our scope might be broadened out through a focus on the social dimensions of synthetic biology and to show how social and cultural ideas influence the content of daily scientific work, we will turn briefly to the use of metaphors and analogies in synthetic biology. This is a topic with potentially profound influences on the development of synthetic biology as a field but that risks being neglected if the focus is on 'implications' or 'impacts' rather than 'dimensions'.

There are several different approaches to synthetic biology (O'Malley, Powell, Davies, & Calvert, 2008), but the one that has become most visible in recent years aims to apply explicit engineering principles and practices to biological systems (Endy, 2005). In their attempts to describe how biology could become more like an engineering discipline, these synthetic biologists draw heavily on metaphors and analogies from many different branches of engineering (including electrical, software and mechanical engineering; see Hellsten & Nerlich, 2011). They borrow from the language of these socially established engineering disciplines to help explain and motivate their attempts to engineer biological systems.

Some of these metaphors and analogies have become so familiar that it is now hard to recognise them as such. For example, the idea of building genetic 'circuits', an analogy borrowed from electrical engineering, is widely used in synthetic biology. And a computational metaphor that often appears is that of 'programming'. Although synthetic biologists are not literally programming cells, the metaphor captures the appropriate spirit and has become an accepted part of the vocabulary of the field.

Another computational term commonly encountered in synthetic biology is 'refactoring'. This refers to the improvement, rationalisation and streamlining of computer software, and it has been imported into synthetic biology to mean improving, rationalising and streamlining naturally existing genes and genomes. Temme, Zhaob, and Voigt (2012) describe their recent work as 'refactoring' the nitrogen fixation gene cluster from the bacteria *Klebsiella oxytoca*. They redesigned the gene cluster *in silico*, removing many of the existing gene regulation mechanisms, feedback loops and redundancies. They then synthesised their simplified, 'refactored' cluster, stripped out the cell's existing version and replaced it with their synthetic one. The refactored cell's nitrogen fixing activity was not improved in this experiment (in fact it was made worse), but it was much easier to understand and manipulate than the wild type, and it had interestingly different characteristics, such as being able to withstand higher levels of certain toxins (Jermy, 2012). This work is regarded as proof of principle that it will be possible in the future to redesign genes and gene clusters for engineering purposes. It also shows how engineering-derived ideas, based on a socially accepted body of pre-existing knowledge, can be imposed on biological systems. It is just one example of many DNA 'rationalisation' activities, the most famous of which is probably the J. Craig Venter Institute's ongoing attempt to identify the minimal genome necessary for life, synthesise it and use it to 'boot up' (a metaphor) a host cell or 'chassis' (an analogy).

A number of the chapters of this book focus on different 'chassis' for synthetic biology. This is a term from mechanical engineering that has been adopted by synthetic biologists to describe the context into which a biological 'circuit' can be inserted. It replaces the more familiar biological word 'cell'. The first use of the term 'chassis' in a biological context is found in Canton (2005), who defines it as "a host cell that is capable of supplying the demands of an engineered system...while simultaneously insulating the engineered system from the environment" (p. 10). In a poster derived from the same work, Canton and Endy (2005) used a picture of an automobile chassis to make their point. They note that there are normally many interactions between a host cell and any engineered system inserted into it, but they argue that an ideal chassis would behave predictably and independently of the synthetic constructs it contains (Canton & Endy, 2005). The central idea is that the chassis will take care of the material needs of the biological system (replication, transcription, translation and degradation, etc.), while the expressed genetic circuit performs the engineer's specified function.

Notably, when a cell is described as a 'chassis', it becomes an object or tool that provides a service to the biological engineer. This raises many interesting questions about the use of this analogy: Do we change our appreciation of the nature of a cell when we call it a 'chassis'? Do we feel freer to manipulate it and treat it as just another tool? Does the word encourage us to forget that we are working with a living thing? Or does the idea of a 'chassis' assist our understanding of what it means to engineer with biology and help to move the field of synthetic biology forward? In summary, what does the word 'chassis' assume? What does it enable? And what does it suppress? Similar questions could be asked for many of the metaphors and analogies used in synthetic biology, and their use is an important social dimension of the field. As a prominent STS researcher observed over 20 years ago, "Power is about whose metaphor brings worlds together" (Star, 1991, p. 52). The metaphors that organise a discipline can prove very consequential. Indeed, it seems that the central metaphors and analogies that are used in the daily workings of synthetic biologists are shaping the types of research questions they are pursuing and thus the types of products and applications that are likely to emerge.

3 FROM SPECULATION TO ANTICIPATION

Synthetic biology is a young field, but many of the broader expectations associated with it can be traced back several decades. For example, in an article written over 40 years ago, the biologist James Danielli talks about his aims for biology in an 'age of synthesis'. His ambitions include "designing bugs for other planets, or making crop plants that such as wheat that could fix their own nitrogen, or even for synthesising micro-organisms that could eventually do the job of the chemical engineering industry" (Chedd, 1971, p.124). There are striking parallels to the hopes we routinely hear expressed for synthetic biology today.

Danielli was talking about the future, and much of the current discussion of synthetic biology is also about a hoped-for future. It is about what synthetic biology *will* deliver rather than what it has already delivered. Many articles and reports about synthetic biology quickly slip into the future tense. And many of these discussions about the future draw on past precedents. For example, when suggesting that the field could herald a new 'industrial revolution' based on biomanufacturing, synthetic biologists often draw parallels with the first industrial revolution of the late eighteenth and early nineteenth centuries and the microprocessor revolution of the late twentieth century (Kitney, 2007).

Although we should be aware that claims like these are speculations about the future, this does not mean they are unimportant or can simply be dismissed. Work in the field of the sociology of expectations has shown that actions in the present are made legitimate by promises about the future (Borup, Brown, Konrad, & Van Lente, 2003). This is because the future has effects in the present: if something is expected to succeed, people will invest in it, in turn making it more likely to succeed. Vision and momentum are vital to get a new field off the ground (Brown, 2003) and, since synthetic biologists are attempting to build a new field, they require buy-in, resources and enthusiasm.

Because the future does play an important role in shaping technologies, it becomes important to draw a distinction between untethered speculation and more grounded anticipation. Almost every week, it is possible to come across news headlines about speculative synthetic biology applications. Sometimes, this is an extension of work that has taken place in laboratories. For example, Pam Silver's group at Harvard Medical School has inserted photosynthetic microbes (cyanobacteria) into transparent zebrafish embryos (Collins, 2012). The hope is that in the future, these fish would be able to get some of their energy from photosynthesis. This is a long way off, but not as far off as a green-skinned human who only needs to go out in the sun to fulfil their nutritional needs (Singer, 2010). More extreme speculations include talk of the 'de-extinction' of species like woolly mammoths and sabre-toothed tigers, using the tools of synthetic biology (Kolata, 2013)—although it is notable that the issue of how these animals might survive without their original habitats, foodstuffs and parasites is often omitted. The idea has even been floated of bringing back Neanderthals, perhaps using a human as a surrogate mother (Friedman, 2013). Such speculative discussions are undoubtedly provocative, but questions arise about how much they actually further our understanding of the social dimensions of synthetic biology, since they often seem to be based on the assumption that these developments are just around the corner and that all the relevant technological advances have already been made (Marris & Rose, 2012).

Our understanding of the social dimensions of synthetic biology is very different when we move from speculations to a closer investigation of what day-to-day scientific work actually involves. Much current research focuses on developing foundational technologies that will underpin the engineering of biological systems, and many synthetic biology laboratories are working on developing tools for engineering biology faster, more predictably and more reliably. For example, researchers in the

California-based BIOFAB (a public-benefit facility for advancing biotechnology) noticed that different combinations of promoters and ribosome binding sites (RBSs) had considerably different effects on protein expression levels in *E. coli* (their chassis of choice). They have worked on engineering more precise control of gene activity by designing promoters and RBSs that do not interfere with the downstream gene, leading to more predictable levels of gene expression (Mutalik et al., 2013).

Synthetic biologists at Imperial College London are also working to develop tools for more reliable engineering. For example, they wanted to circumvent the time-consuming work of transforming and culturing living cells, so they produced a cell-like environment inside a test tube and succeeded in synthesising proteins faster and in much larger quantities than would have been possible using cellular techniques (Chappell, Jensen, & Freemont, 2013). This could allow for the more rapid production of standardised biological 'parts' in the future. The use of engineering analogies is strongly present in the work of researchers at both the BIOFAB and Imperial. For example, in commenting on his work, Chappell noted: "One of the major goals in synthetic biology is to find a way to industrialise our processes so that we can mass-produce these biological factories in much the same way that industries such as car manufacturers mass produce vehicles in a factory line" (Smith, 2013). Here, we see a vision for the future of synthetic biology that embraces the ideas of industrialisation and mass production and requires the development of fast and reliable techniques.

Today's synthetic biology is much better represented by the experimentation currently taking place in these laboratories than by the more speculative ideas discussed earlier. We think that studies of the social dimensions of synthetic biology should focus primarily on the realities of the research being done, as opposed to getting caught up in an imaginary realm. Again, this distinguishes our approach from one that is focused primarily on 'implications'. As the implications of synthetic biology may be located in the distant future, there is a danger that focusing on implications will pick up on the more speculative applications of the technology, on hypothetical rather than actual developments (Nordmann & Rip, 2009). This could involve constructing narratives around the rights of future Neanderthals, for example, rather than addressing the relevance and applicability of using past models of industrialisation to guide current activities in synthetic biology.

Based on this understanding, we advocate a shift in language and thinking from 'speculation' to 'anticipation'. We do think it is important to anticipate the future and to discuss possible developments in synthetic biology (and who they are likely to affect), but to do so in a manner that is grounded in actual and ongoing scientific work. A variety of formal methodologies for doing this type of anticipatory work have been developed, including approaches such as scenario development (Aldrich, Newcomb, & Carlson, 2008), constructive technology assessment (Schot & Rip, 1997), midstream modulation (Fisher et al., 2006) and anticipatory governance (Guston, 2008). At first glance, more grounded discussions might appear to be about 'merely' technical visions of the future (e.g. one where it is possible to 'plug and play' with BIOFAB-designed promoters and RBSs), but in practice, even technically oriented visions of the near future contain value judgements, because they point to a future that is seen as *desirable*, for example, one where the more systematic engineering of biology would be a good thing. It is important to remember that visions are developed by people and that not all visions of the future are universally shared. With this in mind, we should pay careful attention to who is articulating a given vision, what position they are speaking from and how they see synthetic biology and its contribution to the future.

4 FROM PUBLIC ACCEPTANCE TO PUBLIC GOOD

This leads us to questions about who should have a voice in the direction of synthetic biology, a point where discussion often turns to 'public acceptance'. We want to suggest that a more constructive way of thinking about public engagement in this field is to think about how synthetic biology can contribute to the public good.

A recurring problem with discussions that are framed in terms of public acceptance is that they often assume the reason people do not accept new technologies is because they are ignorant about them. The implied solution to this problem is to teach the public more about the science, in the hope that they will become more favourably disposed towards it. The pervasiveness of this argument among scientists and policymakers has led STS researchers to give it the label of the 'deficit model', which "conceptualizes the lay mind as an empty bucket into which the facts of science can and should be poured" (Gregory & Miller, 1998, p. 89). However, several decades of social science research have shown that the relationship between public understanding of science and public support for science is ambiguous at best. Concerted attempts to increase public knowledge have not typically led to greater support for science, and we see that social groups and countries that have greater scientific knowledge do not correlate neatly with those that are more positive about scientific developments (Durant & Legge, 2005; Marks, 2009). Interestingly, this phenomenon is not limited to 'the general public'—scientists themselves sometimes find that the more they know about a particular scientific field, the more aware they become not only of the potential opportunities but also of all the challenges and nuances involved in advancing the field, and the more ambivalent (or even sceptical) their views become about whether and when the fruits of their labour will be realised.

Many discussions of public acceptance also assume that the reasons for the failure of the public to embrace a particular technology revolve around the public's (unfounded) concerns about its risk—either to the environment or to their own personal health. A conclusion that is often drawn is that, to reassure the public, it is necessary to demonstrate that the technology is safe and adequately regulated. These assumptions are frequently made in the absence of any attempt to find out what it is that actually concerns or interests people when they are confronted with a new technology.

One example that illustrates the problem is the debate over genetically modified (GM) food crops in the United Kingdom. In this case, an assumption was made that the key public concerns were about the risks of cross contamination of other plant species and possible physical harm from eating GM crops. As a result, concerted efforts were made to undertake risk assessments about the safety of these crops. But this narrow focus on safety left out a number of other factors that were also important in shaping people's attitudes towards GM. These included the monopolisation of the technology and the control of intellectual property rights by large corporations, the implications of GM for global industrial food distribution and trade, and threats to indigenous food cultures and rural farmer livelihoods (Stirling, 2012). A narrowly framed risk assessment, focusing exclusively on safety, could not address these factors, which then ended up becoming completely excluded from regulatory procedures for decision-making. This example illustrates the dangers of assuming in advance the issues that are likely to be most important to people when they engage with new technologies.

The GM debate is often cited in discussions of synthetic biology in Europe (indeed, it is often the 'elephant in the room'; Marris, 2012), but, as Torgensen and Schmidt (2013) point out, it should not necessarily be assumed that the public discussions around synthetic biology will play out in the same way. Synthetic biology, like GM before it, might become wrapped up in a debate about risks and safety, particularly if similar assumptions are made about public attitudes. But it could be argued that synthetic biology has more in common with technologies such as nanotechnology, which like synthetic biology is an extremely interdisciplinary emerging field that promises to herald a new industrial revolution. Or parallels could be drawn with information technology, which synthetic biology models itself on to a large extent (as we see in discussions of open-source biology and 'reprogramming' DNA). These other comparators could prove influential in the way that public discussions of synthetic biology are framed.

A third problem with the idea of public acceptance is that it assumes that 'the public' is a singular, predefined group that accepts or rejects particular technologies as a whole. In practice, we know there are many different and overlapping groups in the public sphere (including scientists and engineers), who have a range of different interests, concerns and expertise. In synthetic biology, we are also seeing the rise of the do-it-yourself biology movement, or 'DIYbio'. This is an extremely diverse group of amateur scientists, including ex-professionals and moonlighting working scientists, artists using biology as a material, enthusiastic school children and complete novices. Movements like DIYbio, as well as patient advocacy groups and non-governmental organisations (NGOs), challenge the idea that 'science' and the 'public' can be firmly separated. They also challenge the idea that the public should passively 'accept' technological developments. These groups are not passive, they are active and engaged, and they can have profound effects on the development of technologies. To give one well-documented example, AIDS patient activists had a huge impact on the development of clinical trials for AIDS treatment in the 1980s (Epstein, 1995).

In response to criticisms of the deficit model, what we have seen over the past 20 years is a shift in language at the policy level from public acceptance or understanding of science to public engagement/dialogue with science (House of Lords, 2000). Rather than advocating one-way communication from experts to the public, public engagement is meant to involve a two-way dialogue, with genuine deliberation and negotiation around scientific and technological issues. Although nonscientists cannot easily contribute to discussion of the technical details of research, biotechnology in general (and synthetic biology in particular) aims to serve broad social and economic goals. The choice and prioritisation of these goals *can* be opened up to wider public discussion, and, because of their breadth, a discussion of these goals can benefit from input from a variety of perspectives. For example, a discussion of priorities for food security stands to benefit from contributions by molecular biologists, agronomists, economists, farmers, local communities, small businesses and food consumers, rather than being treated as a scientific or technical challenge alone (Nuffield Council on Bioethics, 2012).

There are different forms of public engagement that are more or less appropriate for different circumstances (Delgado, Kjølberg, & Wickson, 2011). For synthetic biology, the most valuable form(s) of public engagement may well be with the presumed end users or beneficiaries of the technology, rather than with a group of 'invited' members of the public who are called in to take part in a focus group or citizens' jury, but have no involvement or direct stake in the technology's development. The latter, it must be said, has been the dominant approach of formal public engagement activities to date in the United Kingdom.

4.1 Synthetic biology for a particular purpose—the arsenic biosensor

At this point, it may be helpful to turn to a specific example, and here, we highlight an application of synthetic biology that aims to benefit those in developing countries. The Wellcome Trust recently funded a research team to develop an arsenic biosensor using synthetic biology (Bailey, 2013). This particular project originated in the work of two undergraduate teams that took part in the annual International Genetically Engineered Machine (iGEM) competition—the University of Edinburgh team from 2006 and the Cambridge University team from 2009—and it is now being taken forward by research laboratories at these two institutions.

Arsenic is a toxic contaminant of ground water in many parts of the world, including the United States and many drinking water wells in South Asia. It can lead to painful lesions on hands and feet, as well as various forms of cancer. It is extremely hard to determine which water wells contain arsenic and which do not, and existing tests either take too long (because they require the water sample to be sent to a centralised laboratory) or they have to be carried out by trained operators (because they produce a toxic gas). But some bacteria are very good at detecting low levels of arsenic—which is where synthetic biology comes in. Bacterial genes involved in arsenic detection can be coupled to genes that produce colour pigments.

The resulting genetic construct can be then be inserted into another bacterium that will change colour if arsenic is present in a well-water sample.

This is the start of a technical solution to the problem of arsenic detection. But the development of any technology that is going to be used by people in a particular social and environmental context involves much more than the invention of a technical device. The Cambridge/Edinburgh research team is currently investigating the potential for their arsenic biosensor for use in Nepal, and in so doing, they are adapting the design of the device. One of the early decisions they made was to put the synthetic construct inside *Bacillus subtilis* rather than the more familiar chassis *E. coli*. This is because *B. subtilis* produces spores, which are much more transportable than *E. coli* and can survive better in hot climates. Although this modification is a technical one, it is a design decision made in response to the particular conditions in which the device will be used.

What is particularly notable about this project is that the research team includes not only two synthetic biologists but also a business advisor with expertise in legal, financial and marketing issues and a development expert who specialises in the study of Nepal and who can engage with the needs and priorities of the local communities. Emphasis has been placed on involving the potential users of the product in the design process from a relatively early stage. These include representatives from government, business, the regulatory sphere and NGOs, as well as the local community. By consulting with these communities, the research team were made aware of their requirements and are adapting the design of the device accordingly. One design consideration that came to the fore was the affordability of the device and the need, ideally, for it to cost less than 30 US cents per test. Another was how to dispose of the genetically modified bacteria safely after the test had been carried out.

But more unexpected issues also emerged. For example, local people said they would prefer a digital readout to a colour one (Bailey, 2013). This is something that could have only have been learnt through dialogue with the potential users of the technology—researchers had assumed that a colour readout would be useful and straightforward. Furthermore, the arsenic biosensor proposal also intersects with context-specific cultural factors, such as the fact that for a poor Nepalese father, having your drinking well marked as being poisoned with arsenic might make it difficult to marry off your daughters because of perceived long-term health risks associated with drinking contaminated water. We see here that the detection of arsenic can become a socially sensitive issue. What this shows is that it is not just a case of developing a synthetic biology device and handing it over to a user but that the process of developing a technology that works in a particular context is far more complex.

Although underlying problems of water infrastructure in parts of South Asia remain, the arsenic biosensor is a useful example for two reasons. Firstly, it appears to provide a solution to a problem where the existing or alternative solutions do not work to the satisfaction of the affected community. Rather than being a technology in search of a problem (like a hammer in search of a nail), this example seems to represent a problem that might be usefully be addressed using a synthetic biology approach. Secondly, this case study is a good example of an attempt to incorporate environmental, political and societal considerations into synthetic biology design, by engaging with relevant experts and stakeholders from an early stage. This makes it more likely that the technology will eventually be adopted by the groups it has been designed for and that it will contribute to the public good.

4.2 The public good

In this section, so far, we have argued for a shift away from notions of public acceptance towards the idea of public good. But what exactly is meant by 'the public good'? The answer is not a straightforward one, because there is no single, unchanging notion of the public good; it requires ongoing negotiation and context-dependent discussion (Nuffield Council on Bioethics, 2012). This is why the idea of the public good is so closely linked to public and stakeholder dialogue, because such dialogue provides a way of bringing in a range of voices who can contribute to our understanding of the public good in particular circumstances. It is likely that different groups will disagree, but being aware of contrasting perspectives is an essential part of enriching the discussion and shaping priorities.

A series of questions can be raised to help arrive at an understanding of the public good in a particular context. These questions include the following: What is the purpose of introducing the technology? Who will benefit from it? Will the introduction of the technology have clear social benefits, as well as economic ones? Is this just a technology in search of a problem, or one that is fulfilling a genuine need? Could the same job be done more quickly, cheaply or efficiently by using an alternative technology or by turning to social or economic interventions?

The introduction of values or principles can also be useful in discussions of the public good, although trade-offs may prove necessary. Such values might include sustainability (Wiek, Guston, Frow, & Calvert, 2012), biodiversity conservation (Redford, Adams, & Mace, 2013), justice (Reardon, 2013), responsibility (Owen, Macnaghten, & Stilgoe, 2012), solidarity (Prainsack & Buyx, 2011) and human flourishing (Rabinow & Bennett, 2012). Like the public good, what these values mean is also the subject of ongoing discussion and debate. Some synthetic biologists have explicitly incorporated values into their mission statements. For example, the BioBricks Foundation, one of the most important institutions driving parts-based approaches to synthetic biology, has the tagline 'biotechnology in the public interest' and says that it aims "to benefit all people and the planet" (biobricks.org). This is clearly a broad and ambitious aim (it is hard to think of any technological development that would benefit *all* people *and* the planet), but the important point about the idea of 'biotechnology in the public interest' is that the public interest is not one that any particular group of scientists, policymakers or NGOs can decide in isolation—it is something that needs to be debated among a broad range of groups.

5 FROM REGULATION TO GOVERNANCE

This talk of involving multiple stakeholders in the pursuit of the public good leads us to the topic of 'governance' more broadly and takes us to the fourth and final shift we are advocating in our overview of the social dimensions synthetic biology: from regulation to governance. We think this shift is productive because there is a tendency to assume that many of the 'issues' raised by synthetic biology can be dealt with through formal regulation. There is much debate about whether new, synthetic biology-specific regulations are needed or whether existing mechanisms for dealing with genetically modified processes and products are adequate. But many of these discussions miss the point: instead of thinking in terms of regulation, it is often more useful to think in terms of governance.

The language of 'governance' started to be adopted in the early 2000s. It marked a general shift away from focusing on top-down, expert-driven approaches to regulation and policymaking towards advocating bottom-up approaches involving a greater diversity of interested groups (Murphy & Levidow, 2006). Governance involves forms of decision-making that aim to be inclusive and deliberative, rather than hierarchical. It is much broader than regulation and not only includes the passing of laws but also encompasses institutions such as the patent system, research funders, biosafety panels, institutional ethics committees and advisory bodies that all shape the ways in which science and technology develop.

In the context of synthetic biology, thinking in terms of governance draws our attention to the many different groups that currently play a role in the field (Zhang et al., 2011). To give a snapshot, this includes interdisciplinary research centres like the Synthetic Biology Engineering Research Centre (which brings together five universities in the United States); not-for-profit institutes like the J. Craig Venter Institute and the Joint BioEnergy Institute; funding agencies like the European Commission, the Biotechnology and Biological Sciences Research Council and the Engineering and Physical Sciences Research Council in the United Kingdom and the National Science Foundation and the Defense Advanced Research Projects Agency in the United States; small companies like Life Technologies, Synthetic Genomics and Amyris Biotechnologies; larger companies like Shell and GlaxoSmithKline; governmental advisory bodies like the Technology Strategy Board in the United Kingdom and the Presidential Commission for the Study of Bioethical Issues in the United States; and NGOs such as the ETC Group. Synthetic biology governance is also influenced by groups that are hard to categorise and are not expert-led, such as the DIYbio community and the iGEM undergraduate competition. And we should not forget that the field of synthetic biology is very interdisciplinary, bringing together many different kinds of scientists and engineers as well as lawyers, bioethicists, philosophers, economists, anthropologists, sociologists and even artists and designers. All these groups can and do play a role in the governance of synthetic biology.

5.1 Responsible research and innovation

One approach to governance that is becoming prominent in discussions of synthetic biology in the United Kingdom and Europe is Responsible Research and Innovation (RRI). We focus on it here as it fits nicely with our emphasis on the 'dimensions' rather than the 'implications' of synthetic biology. RRI encourages a focus on the entire process of innovation, rather than primarily being concerned with applications and impacts. The idea of 'responsible' research and innovation also ties in with the discussion earlier about the role of values in negotiations about the public good.

RRI starts from the recognition that risk regulation is an inadequate tool for dealing with the governance of emerging technologies. This is because, when technologies are still emerging, it is not possible to calculate their risks in advance nor is it possible to predict their unintended and unforeseen consequences. The famously incorrect prediction attributed to the founder of IBM in 1943 that there would be a total world market for approximately five personal computers makes this point (Weiss, 2007). Because precise prediction is not possible with emerging technologies, it becomes important for governance mechanisms to attempt to make allowances for uncertainty. One way of doing this is to try to encourage flexibility, resilience and diversity in technological development, in the hope of avoiding lock-in to a limited range of paths that later turn out to be undesirable.

Another problem with focusing on regulating risk is that it gives the impression that the aim is to limit, stop or prevent the development of a particular technology. However, the point of RRI is that rather than focusing discussion around what we *do not* want the technology to do, the aim is think about what we *do* want it to do and "how the targets for innovation can be identified in an ethical, inclusive, democratic and equitable manner" (Owen et al., 2012, p. 754). This involves shifting the discussion away from risk to focus on the visions, aims and intentions that drive scientific and technological development. Governance in RRI puts an emphasis on the importance of reflection on the purposes and motivations for scientific and technological developments and provides an opportunity for inclusive deliberation about the direction of technological innovations, in this way "opening up opportunities for these to be directed towards socially desirable ends" (Owen et al., 2012, p. 754). This may, for example, involve public and stakeholder involvement in the formulation of 'grand societal challenges' at research council level.

RRI can be defined as "collective care for the future through the stewardship of innovation in the present" (Jirotka, Eden, & Stahl, 2012, p. 10). Owen et al. (2012) put forward four key features of RRI: it should be anticipatory, reflective, inclusive and responsive. The aim to be anticipatory ties into the discussion earlier about the shift from speculation to anticipation. The aim of anticipation is not to predict the future, but to explore *plausible* future pathways for the technology. This requires thinking through various possibilities, considering possible intended and unintended consequences and acknowledging uncertainty. Being reflective demands reflecting on commitments and promises about the impacts and applications of the technology;

being inclusive involves opening up this reflection on technology to a range of stakeholders, users and wider publics (something that has been part of the innovation process in new product development for many decades); and finally, for innovation to be responsive, it should be iterative and flexible and keep options open by encouraging diversity. As can be seen in the brief description of these four requirements, responsibility in the context of RRI is not necessarily the kind of responsibility that has to be taken at an individual level by a scientist working in synthetic biology. Patenting systems, for example, can have significant effects on questions about the ownership, distribution and eventual public good of a technology. We have seen that there is a range of different people and organisations involved in innovation in synthetic biology, and responsibility for responsible innovation should be distributed and shared among this whole network.

The RRI approach to governance stresses the importance of interdisciplinary collaborations, because conversations across disciplines, including those between the natural and social sciences, can help all those involved think in new, more anticipatory, reflective, inclusive and responsive ways about their work. In this sense, RRI may involve creating new interdisciplinary spaces and interactions (Reardon, 2013), where people can reflect together on the directions that a technology is taking and ask whether it is helping build a future that contributes to the public good.

CONCLUSIONS

Much of the current work in synthetic biology is concerned with establishing foundational tools and technologies, as a platform upon which to innovate with biology. In a similar vein, we have used this chapter to outline some of the foundational ideas that underpin our approach to the social study of synthetic biology. The shifts we have outlined represent broad trends in thinking that have arisen through the past couple of decades of social science research. The shifts from implications to dimensions, from speculation to anticipation, from public acceptance to public good, and from regulation to governance, are each important aspects of how we approach the study of the social dimensions of synthetic biology. This type of background or contextual thinking is often left out when social scientists relate their research to synthetic biologists, and we thought it was worth devoting some time and space to an explicit articulation of these underpinning ideas. Together, they lay the groundwork upon which more specific issues in synthetic biology might be addressed—in much the same way as the foundational technologies of reading and writing DNA allow specific and targeted applications to be developed.

These broad ideas shape our thinking when it comes to many of the more specific 'ELSI' issues that crop up in the context of synthetic biology. For example, when considering biosecurity in synthetic biology, we start not from the perspective of thinking immediately about how top-down regulatory approaches could help to mitigate biosecurity threats. Nor do we jump straight into speculations about future biosecurity risks of synthetic biology. Rather, looking at the issue from the perspective of governance, we might start by mapping the landscape of actors and their relationships and considering how their (changing) practices are contributing to what is described as an emerging threat. Understanding how knowledge and materials flow among DNA synthesis providers, community laboratories and synthetic biology research institutes might reveal opportunities to make small modifications to current practices that would reduce biosecurity concerns and foster a culture of responsible practice—modifications that may not require formal regulatory action.

Or to pick up on another example, when it comes to making sense of thorny questions about 'creating life', we do our best to cultivate an awareness of, and sensitivity to, the diversity of views that might be held among various groups—be they lay citizens, activists, policymakers, synthetic biologists and so on. And we work to understand the reasons behind the claims they make, to try and ascertain what is at stake when debates emerge. From such a vantage point, we might be able to identify methods for fostering discussion that allow values and expectations to be more explicitly acknowledged. We may find, for example, that some groups' concerns about 'creating life' might revolve around issues to do with the inherently unpredictable nature of emerging technologies and the difficulties of control (both biological and regulatory) when dealing with living things. Bringing issues such as these to the surface makes the discussion more tractable and allows it to become more focused and relevant.

Most fundamentally, we are interested in the nature of the social contract between science and modern democratic societies and how this contract might be written and calibrated in such a way to allow both to flourish. And we hope that the ideas presented earlier might serve as useful tools for starting discussion, although we realise that they may be rather different from the tools and protocols presented in the other chapters of this volume.

References

Aldrich, S., Newcomb, J., & Carlson, R. (2008). Scenarios for the future of synthetic biology. *Industrial Biotechnology, 4*(1), 39–49.

Bailey, P. (2013). Feature: The biggest poisoning in history. *Wellcome Trust*, Available at: http://www.wellcome.ac.uk/News/2013/Features/WTP051176.htm.

Borup, M., Brown, N., Konrad, K., & Van Lente, H. (2003). The sociology of expectations in S&T. *Technology Analysis & Strategic Management, 18*(3/4), 285–298.

Brown, N. (2003). Hope against hype—Accountability in biopasts, presents and futures. *Science Studies, 16*(2), 3–21.

Canton, B. (2005). Engineering the interface between cellular chassis and integrated biological systems: Thesis proposal. Available at: http://dspace.mit.edu/bitstream/handle/1721.1/19813/BC.ThesisProposal.pdf?sequence=1.

Canton, B., & Endy, D. (2005). Engineering the interface between cellular chassis and integrated biological systems. In *Poster presented at the sixth international conference on systems biology, Boston, MA, 19–24th October 2005*, Available at: http://dspace.mit.edu/bitstream/handle/1721.1/29802/BC.ICSB05.pdf?sequence=1.

Chappell, J., Jensen, K., & Freemont, P. S. (2013). Validation of an entirely in vitro approach for rapid prototyping of DNA regulatory elements for synthetic biology. *Nucleic Acids Research*, *41*, 3471–3481. http://dx.doi.org/10.1093/nar/gkt052.

Chedd, G. (1971). Danielli the Prophet. *New Scientist and Science Journal*, (21 January 1971), 124–125.

Collins, J. (2012). Bits and pieces come to life. Scientists are combining biology and engineering to change the world. *Nature*, *483*, S08–S10.

Delgado, A., Kjølberg, K. L., & Wickson, F. (2011). Public engagement coming of age: From theory to practice in STS encounters with nanotechnology. *Public Understanding of Science*, *20*(6), 826–845.

Durant, R. F., & Legge, J. S.Jr., (2005). Public opinion, risk perceptions, and genetically modified food regulatory policy: Reassessing the calculus of dissent among European citizens. *European Union Politics*, *6*, 181–200.

Elowitz, M. B., & Leibler, S. (2000). A synthetic oscillatory network of transcriptional regulators. *Nature*, *403*(6767), 335–338.

Endy, D. (2005). Foundations for engineering biology. *Nature*, *438*, 449–453.

Epstein, S. (1995). The construction of lay expertise: AIDS Activism and the forging of credibility in the reform of clinical trials. *Science, Technology, & Human Values*, *20*(4), 408–437.

ETC Group, (2007). *Extreme genetic engineering: An introduction to synthetic biology*. Available at: http://www.etcgroup.org/en/materials/publications.html?pub_id=602.

Fisher, E. (2005). Lessons learned from the Ethical, Legal and Social Implications program (ELSI): Planning societal implications research for the National Nanotechnology Program. *Technology in Society*, *27*, 321–328.

Fisher, E., Mitcham, C., & Mahajan, R. (2006). Midstream modulation of technology: governance from within. *Bulletin of Science, Technology and Society*, *26*, 485–496.

Friedman, R. (2013). Interview with George Church: Can Neanderthals Be Brought Back from the Dead? *Spiegel*, 18th January 2013. Available at: http://www.spiegel.de/international/zeitgeist/george-church-explains-how-dna-will-be-construction-material-of-the-future-a-877634.html.

Gardner, T. S., Cantor, C. R., & Collins, J. J. (2000). Construction of a genetic toggle switch in *Escherichia coli*. *Nature*, *403*, 339–342.

Gregory, J., & Miller, S. (1998). *Science in public: Communication, culture and credibility*. New York: Plenum Press.

Guston, G. (2008). Innovation policy: Not just a jumbo shrimp. *Nature*, *454*, 940–941.

Hellsten, I., & Nerlich, B. (2011). Synthetic biology: Building the language for a new science brick by metaphorical brick. *New Genetics and Society*, *30*(4), 399–413.

House of Lords. (2000). *Science and society*. House of Lords Select Committee on Science and Technology, third report. Available at: http://www.publications.parliament.uk/pa/ld199900/ldselect/ldsctech/38/3801.htm.

Jermy, A. (2012). Synthetic biology: We can rebuild you. *Nature Reviews Microbiology*, *10*, 378. http://dx.doi.org/10.1038/nrmicro2803.

Jirotka, M., Eden, G., & Stahl, B. (2012). The challenges of responsible research and innovation in contemporary ICT research. Presentation at ESRC Research Methods Festival, St Catherine's College Oxford, 2nd–5th July 2012. Available at: http://eprints.ncrm.ac.uk/2807/1/NCRMJirotkaUpload.ppt.

Kitney, R. (2007). Synthetic biology: Engineering biologically-based devices and systems. *IFMBE Proceedings*, *16*, 1138–1139.

Kolata, G. (2013). So you're extinct? Scientists have gleam in eye. *New York Times*, 18th March 2013. Available at: http://www.nytimes.com/2013/03/19/science/earth/research-to-bring-back-extinct-frog-points-to-new-path-and-quandaries.html?pagewanted=all&_r=0.

Marks, N. (2009). Public understanding of genetics: The deficit model. In *Encyclopaedia of life sciences*. Chichester: John Wiley and Sons.

Marris, C. (2012). The elephant in the room: We must avoid another GM. Presentation at Social Scientists Adventures in Synthetic Biology, 19th June 2012. Available at: http://www.kcl.ac.uk/sspp/departments/sshm/research/csynbi/ProgrammeSocialScientistsAdventuresSB.pdf.

Marris, C., & Rose, N. (2012). Let's get real on synthetic biology. *New Scientist*, *214*(2868), 28–29.

Murphy, J., & Levidow, L. (2006). *Governing the transatlantic conflict over agricultural biotechnology: Contending coalitions, trade liberalisation and standard setting*. London: Routledge.

Mutalik, V. M., Guimaraes, J. C., Cambray, G., Lam, C., Christoffersen, M. J., Mai, Q.-A., et al. (2013). Precise and reliable gene expression via standard transcription and translation initiation elements. *Nature Methods*, *10*(4), 354–360.

Nordmann, A., & Rip, A. (2009). Mind the gap revisited. *Nature Nanotechnology*, *4*, 273–274.

Nuffield Council on Bioethics, (2012). *Emerging biotechnologies: Technology, choice and the public good*. London: Nuffield Council on Bioethics, Online at: http://www.nuffieldbioethics.org/emerging-biotechnologies.

O'Malley, M., Powell, A., Davies, J., & Calvert, J. (2008). Knowledge-making distinctions in synthetic biology. *BioEssays*, *30*(1), 57–65.

Owen, R., Macnaghten, P., & Stilgoe, J. (2012). Responsible research and innovation: From science in society to science for society, with society. *Science and Public Policy*, *39*, 751–760.

Prainsack, B., & Buyx, A. (2011). *Solidarity: Reflections on an emerging concept in bioethics*. London: Nuffield Council on Bioethics.

Rabinow, P., & Bennett, G. (2012). *Designing human practices*. Chicago: University of Chicago Press.

Reardon, J. (2013). On the emergence of science and justice. *Science Technology and Human Values*, *38*(2), 176–200.

Redford, K. H., Adams, W., & Mace, G. M. (2013). Synthetic biology and conservation of nature: Wicked problems and wicked solution. *PLOS Biology*, *11*(4), e1001530.

Royal Academy of Engineering, (2009). *Synthetic biology: Scope, applications and implications*. London: Royal Academy of Engineering, Online at: http://www.raeng.org.uk/news/publications/list/reports/Synthetic_biology.pdf.

Schot, J., & Rip, A. (1997). The past and future of constructive technology assessment. *Technological Forecasting and Social Change*, *54*, 251–268.

Singer, E. (2010). Photosynthetic fish and other oddities. *MIT Technology Review*, Available at: http://www.technologyreview.com/view/418802/photosynthetic-fish-and-other-oddities/.

Smith, C. (2013). Discovery in synthetic biology a step closer to new industrial revolution. Imperial College London, 1st February 2013. Available at: http://www3.imperial.ac.uk/newsandeventspggrp/imperialcollege/newssummary/news_31-1-2013-12-18-1.

Star, S. L. (1991). Power, Technology and the Phenomenology of Conventions: On Being Allergic to Onions. In J. Law (Ed.), *A Sociology of Monsters: Essays on Power, Technology and Domination* (pp. 26–56). London and New York: Routledge.

Stirling, A. (2012). Opening up the politics of knowledge and power in bioscience. *PLOS Biology*, *10*(1), e1001233. http://dx.doi.org/10.1371/journal.pbio.1001233.

Temme, K., Zhaob, D., & Voigt, C. A. (2012). Refactoring the nitrogen fixation gene cluster from *Klebsiella oxytoca*. *PNAS*, *109*(18), 7085–7090.

Torgensen, H., & Schmidt, M. (2013). Frames and comparators: How might a debate on synthetic biology evolve? *Futures*, *48*, 44–54, Online first http://dx.doi.org/10.1016/j.futures.2013.02.002.

Weiss, A. (2007). Computing in the clouds. *NetWorker*, *11*(4), 16–25.

Wiek, A., Guston, D., Frow, E., & Calvert, J. (2012). Sustainability and anticipatory governance in synthetic biology. *International Journal of Social Ecology and Sustainable Development*, *3*(2), 25–38.

Zhang, J.Y., Marris, C., & Rose, N. (2011). *The transnational governance of synthetic biology: Scientific uncertainty, cross-borderness and the art of governance*. BIOS Working Paper No. 4, London: London School of Economics and Political Science, online at: http://royalsociety.org/uploadedFiles/Royal_Society_Content/policy/publications/2011/4294977685.pdf.

CHAPTER 4

Bacillus subtilis: Model Gram-Positive Synthetic Biology Chassis

Colin R. Harwood[1], Susanne Pohl, Wendy Smith, Anil Wipat

Centre for Synthetic Biology and Bioexploitation, Newcastle University, Newcastle upon Tyne, United Kingdom
[1]*Corresponding author. e-mail address: colin.harwood@ncl.ac.uk*

1 INTRODUCTION

Bacillus subtilis, a low %G+C, Gram-positive, endospore-forming member of the bacterial phylum Firmicutes, is found predominately in the soil and in association with plants. *B. subtilis* is the type species for the genus *Bacillus*, and, following the discovery (Spizizen, 1958) that strain 168 exhibited natural genetic competence, this bacterium has been developed as a high tractable model for Gram-positive bacteria and for the study of basic metabolic and cellular differentiation processes such as sporulation, genetic competence and biofilm formation. The accumulation, over more than half a century, of knowledge of the biochemistry, genetics and physiology of *B. subtilis* has been enhanced in recent years by a number of systematic 'omics' analyses. As a result, *B. subtilis* is one of the most intensively studied and genetically amenable microorganisms and a suitable chassis for a wide range of synthetic biology applications.

B. subtilis is able to grow both in nutrient media and in chemically defined salt media in which glucose, malate and other simple sugars provide sources of carbon and ammonium salts or certain amino acids as sources of nitrogen (Harwood & Archibald, 1990). *B. subtilis* strain 168, on which most studies are performed, is a tryptophan auxotroph (*trpC2*) and therefore requires the addition of tryptophan to the growth media, even those containing acid-hydrolysed proteins such as casein. An analysis of the origins of *B. subtilis* 168 indicates that it was a derived from *B. subtilis* Marburg (ATCC 6051^T), the type strain of both *B. subtilis* and *B. subtilis* subsp. *subtilis* (Zeigler et al., 2008). More recently, a prototrophic variant of strain 168, called BSB1, has been isolated by transformation with DNA from strain W23 (Nicolas et al., 2012). BSB1 is increasingly replacing strain 168 in more systematic research programmes.

Strains of *B. subtilis* are available from a variety of international culture collections including the *Bacillus* Genetic Stock Center (http://www.bgsc.org), which has an extensive collection of wild-type and mutant *B. subtilis* strains, bacteriophages and cloning vectors and the Japanese National BioResource Project (www.shigen.nig.ac.jp/bsub).

When subject to nutrient or physical stress, *B. subtilis* initiates a series morphological and physiological responses in what has been described as a bet-hedging strategy (Veening et al., 2008). These include the induction of macromolecular hydrolases (e.g. proteases and carbohydrases), chemotaxis and motility and competence. If these responses fail to reestablish growth, sporulation is induced in portion (typically $\leq 10\%$) of the population, while another part is condemned to cell lysis (cannibalism) to provide the nutrients required to sustain sporulation (Gonzalez-Pastor, Hobbs, & Losick, 2007). Approximately 5% of the *B. subtilis* genome is devoted to the processes of sporulation and germination and its ability to differentiate into highly resistant endospores provides an enormous competitive advantage in environments such as the soil, where long periods of drought and nutrient deprivation are common.

2 THE *B. SUBTILIS* GENOME
2.1 General features

The genome of *B. subtilis* strain 168 was first sequenced in June 1997 by a joint European/Japanese consortium. The resulting sequence was the first for both a Gram-positive and a differentiating bacterium (Kunst et al., 1997). The initial annotation was carried out primarily at the Institut Pasteur (Mozser, 2002) and was facilitated by the relative ease with which ribosome binding sites (RBSs) and variations on the consensus AAGGAGGT located 4–13-base upstream of the ATG start codon could be identified. The genome was estimated to be 4,214,630 bp in length, comprising 4106 protein-encoding genes (a figure now revised; see later), 86 tRNA genes, 30 rRNA genes and 3 stable RNA genes. These are organised into $\sim$1500 operons that are controlled by some 200 regulatory proteins (Ishii, Ki, Terai, Fujita, & Nakai, 2001). Analysis of the *B. subtilis* chromosome revealed high-level organisation and relationships (Rocha, Danchin, & Viari, 1999; Rocha et al., 2001); for example, highly expressed genes and genes expressed during the active growth phase tend to be oriented in the direction of the replication fork, presumably to reduce conflicts between replication and transcription (Kunst et al., 1997). Approximately 5% of the chromosome has been acquired by horizontal gene transfer (Rocha et al., 2001), including the prophage SPβ, defective prophages PBSX and the *skin* element and seven additional prophage-like elements (Kunst et al., 1997). Just over half of the genome is required for cell processes, intermediary metabolism and macromolecular synthesis, while a significant proportion of the remaining genome is required for growth and survival in the environment. To this end, *B. subtilis* can utilise a wide range of substrates and analysis of the genome reveals the presence of a large number of transporter proteins for the substrate uptake (Saier et al., 2002a, 2002b). *B. subtilis* senses the environment *via* 34 two-component signal transducers, one of which (WalRK) is essential for growth (Howell et al., 2003).

Highly expressed genes are typically found close to the origin of replication, where, at high growth rates, a gene dosage effect is seen. In contrast, prophages and genes that have been acquired by horizontal transfer are typically found near the replication terminus where gene dosage effects are minimal. The different replication modes of the leading and lagging strands result in differences in mutational bias such that the distribution of C and G nucleotides on the two strands is inverted at the replication terminus. This leads to an overrepresentation of valine, glutamic acid, arginine and aspartic acid on the leading strand and of serine, phenylalanine, leucine, isoleucine and histidine on the lagging strand (Rocha et al., 2001).

2.2 Genome annotation

The genome of *B. subtilis* strain 168 has recently been resequenced and reannotated (Barbe et al., 2009), making use of rapid and accurate sequencing techniques and high-level annotation platforms. A key element of this annotation is the drawing of a clear distinction between genes of the paleome (required for sustaining and perpetuating growth) and the cenome (required for adaptation to specific niches). Analysis of the cenome strongly suggests that *B. subtilis* is an epiphyte with a significant proportion of the genome encoding proteins required for their interaction with plants and utilisation of plant detritus (Wipat & Harwood, 1999; Barbe et al., 2009). In addition to its ability to metabolise plant-related carbohydrates such as malate and starch, *B. subtilis* exhibits swarming and biofilm activities that facilitate their association with plant surfaces, particularly in the rhizosphere.

The revised *B. subtilis* 168 genome sequence identified a large number of single-nucleotide polymorphisms and insertions and deletions, reflecting the fact that the original sequence was derived from more than 30 individual laboratories, using first-generation sequencing technology. One result was the annotation of 171 new genes, including some pseudogenes, gene remnants and genes resulting from gene fusions and fissions, bringing the number of protein-coding sequences to 4224 (Barbe et al., 2009). One hundred and seven genes of unknown function, originally designated '*y*' genes, were given biologically significant names, on the basis of experimental evidence found in the literature. The updated annotation can be viewed via GenoList (http://genodb.pasteur.fr/cgi-bin/WebObjects/GenoList).

3 GENOME MANAGEMENT AND ANALYSIS OF GENE FUNCTION
3.1 Gene transfer and recombination

Genes can be transferred into *B. subtilis* using a variety of techniques, including phage-mediated transduction, transformation (natural, electro and protoplast) and conjugation. Phage-mediated transduction was widely used for genome mapping and for the introduction of DNA in *B. subtilis* (Errington, 1984; Errington & Jones, 1987). However, nowadays, transduction is largely superseded by transformation and in some cases conjugation systems and, as a result, is limited to very specific purposes (Kearns & Losick, 2003; Verhamme, Kiley, & Stanley-Wall, 2007). Imported homologous DNA is very efficiently recombined with its cognate sequences on the chromosome, a feature that was used extensively for mapping the genome in the pregenomic era and that nowadays facilitates precise genome engineering.

3.2 Transformation

The ability to introduce isolated DNA (e.g. plasmid, chromosomal or PCR amplicons) into *B. subtilis* and, when required, have it recombine with sequences in the chromosome is at the heart of the extraordinary genetic amenability of *B. subtilis*. The most widely used method for introducing DNA into *B. subtilis* is natural transformation, a process first discovered in 1958 (Spizizen, 1958). While electrotransformation and protoplast transformation are possible, both result in low transformation efficiencies (Chang & Cohen, 1979; Bron, 1990). Although protoplast transformation is time-consuming, it is a useful technique for undomesticated strains that exhibit little or no natural transformation competence (Claessen et al., 2008).

Competence development is one of several post-exponential-phase phenomena associated with this bacterium (e.g. peptide antibiotic synthesis, motility and chemotaxis, hydrolytic enzyme production and sporulation). Competence develops naturally during or shortly after the transition from exponential to stationary phase and is the result of a complex regulatory control circuit that, in part, involves an oligopeptide-base quorum-sensing mechanism (Dubnau & Lovett, 2002; Claverys, Martin, & Polard, 2009). Maximally, only about 10% of the cells in a population become competent. The size of DNA fragments that can be taken up is about 20–30 kb (Dubnau, 1993). When the transforming chromosomal DNA is saturating (>1 µg/ml of competent cells), transformation frequencies of up to 5% of the cells can be achieved with homologous chromosomal DNA. However, under these conditions, individual competent cells can take up more than one molecule of DNA and the cotransfer of unlinked genetic markers is possible, a phenomenon referred to as congression. Although this was an issue when transformation was used for genetic mapping, nowadays, congression can be for the introduction of nonselectable genes into the chromosome since $\sim$1% of recombinants will also contain the required nonselected gene.

The transformation of plasmid DNA into *B. subtilis* by natural transformation is possible, although the frequency at which the plasmid becomes established is usually lower than for chromosomal DNA. This is because transforming DNA is converted into a single-stranded form and randomly fragmented during entry. Consequently, only multimeric plasmid DNA (present in most plasmid preparations) or monomers containing the internal repeats required for recircularisation are effective in plasmid-mediated transformation. The situation is rather different for integration plasmids since they do not have to be reassembled into autonomously replicating forms, and instead, the relevant genes are 'rescued' by integration at the targeted insertion site (see later).

While it is possible to clone genes into a suitable vector and to transform directly into a *B. subtilis* host, the efficiency is often too low to recover transformants and *Escherichia coli* is used as an intermediate host. However, it should be mentioned that some genes and/or their products are toxic to *E. coli*.

Competence is best documented for *B. subtilis* strain 168, while undomesticated strains are often more refractile to competence induction and efficient transformation. To counter this, Nijland, Burgess, Errington, and Veening (2010) used protoplast transformation to introduce plasmid pLK into strain NCIB3610, a wild-type strain often used to study phenomena such as biofilm formation. pLK is a derivative of the unstable multicopy plasmid, pLOSS* (Claessen et al., 2008), carrying *comK*, the product of which is the master regulator of competence. The resulting strain (i.e. NCIB3610 (pLK)) was then used as an intermediate for the genetic manipulation. Induction of *comK* from an Isopropyl β-D-1-thiogalactopyranoside (IPTG)-inducible promoter induced competence and increased the transformability of NCIB3610 to chromosomal and plasmid DNA up to 10-fold, even in LB medium that does not normally support efficient competence development. Because ComK has widespread pleiotropic effects on gene expression, after successful transformation of the host, the unstable pLK plasmid can be easily removed and the resulting transformed strain returned to its wild state. Numerous natural transformation protocols have been published (Bron, 1990; Vojcic, Despotovic, Martinez, Maurer, & Schwaneberg, 2012).

3.3 Conjugation

Interspecies conjugation provides a useful alternative means of gene transfer for strains that exhibit little or no transformability. pLS20, a 55 kbp plasmid isolated from a strain of *B. subtilis* natto, is able to mediate conjugation between a variety of *Bacillus* species, including *B. anthracis*, *B. cereus*, *B. licheniformis*, *B. megaterium*, *B. pumilus*, *B. subtilis* and *B. thuringiensis* (Koehler & Thorne, 1987). pLS20 is not itself a useful cloning vector because of its size and absence of a selectable marker; however, it has been used to mobilise the *B. cereus*-derived vector pBC16, a 4 kbp tetracycline-resistance plasmid that acts as the cloning vector.

3.4 Plasmid-based vector systems

Many plasmids that are derived from Gram-positive bacteria use the rolling-circle mode of replication, which is characterised by the uncoupled synthesis of leading and lagging strands from separate origins of replication: the single-strand origin and the double-strand origin (DSO). This results in the formation of single-stranded DNA intermediates that are subsequently converted to double-stranded DNA. DSOs are often active in only a limited number of strains and, although dispensable, this affects the efficiency of replication and plasmid stability. The copy number of rolling-circle plasmids in *B. subtilis* can vary from about 5 to 200 per chromosome.

A number of so-called theta-replicating plasmids have been developed as vectors for *B. subtilis*, including the enterococcal plasmid pAMβ1 (Bruand, Ehrlich, &

Janniere, 1991; Janniere, Gruss, & Ehrlich, 1993) and pWV01 (Leenhouts & Venema, 1993) and the endogenous *B. subtilis* plasmids pLS20 (Koehler & Thorne, 1987; Meijer, de Boer, van Tongeren, Venema, & Bron, 1995) and pBS72 (Titok et al., 2003). Derivatives of the enterococcal plasmid pAMB1 have been used as the basis of a series of vectors for *B. subtilis* (Janniere, Bruand, & Ehrlich, 1990). For example, pHV1431, a *B. subtilis*/*E. coli* shuttle plasmid, carries the pBR322 replication functions for maintenance in *E. coli* and the pAMB1 replication functions for *B. subtilis*. This plasmid has a copy number of 200 in *B. subtilis* and long inserts are generally maintained stably. pWVO1, from *Lactococcus lactis*, is a small (2178 bp) cryptic rolling-circle plasmid (Leenhouts & Venema, 1993) that replicates stably in *B. subtilis*, a variety of other Gram-positive bacteria and even in *E. coli*. pWVO1, which has a copy number of about 5 in *B. subtilis* but 50–100 in *E. coli*, has also been used to develop a number of special-purpose vectors. The plasmid-encoded RepA function of this plasmid can be provided in *trans*, which means that, with the provision of a suitable *E. coli* helper strain, a RepA$^-$ version of this plasmid can provide the basis of a *B. subtilis* integration system (Leenhouts, 1995; Leenhouts et al., 1996).

The general lack of native *B. subtilis* antibiotic-resistance plasmids means that *B. subtilis* cloning vectors were initially developed using plasmids from other Gram-positive bacteria, such as *Staphylococcus aureus* and *L. lactis* (Bron, 1990; Janniere et al., 1993). The replication functions and/or antibiotic-resistance genes of *S. aureus* plasmids pUB110, pC194 and pE194 (Ehrlich, 1977; Gryczan & Dubnau, 1978) are still in common use, even though vectors based on endogenous *Bacillus* plasmids have now been developed.

3.5 Special-purpose vectors

For many purposes, stable cloning into *B. subtilis* is best achieved using an integrative vector. Integration vectors can be used to generate knockout and fusion mutants, to provide a stable platform for expressing genes (e.g. in complementation studies), for inserting reporter gene constructs and for the study of gene function and essential genes. Integration vectors are plasmids with DNA sequences that are homologous to the *B. subtilis* chromosome but that replicate in *E. coli* and not in *B. subtilis*. Therefore, depending on their construction, the entire vector or a part of it, will integrate into the chromosome of the host by homologous recombination. Two types of recombination are responsible for the integration event, single-crossover recombination events and double-crossover recombination events.

In single-crossover recombination, often referred to as 'Campbell-type' integration, multicopy plasmids based on pBR322 are typically used. However, if the plasmid carries *Bacillus* DNA sequences that are toxic to *E. coli*, low-copy-number plasmids based on pSC101 are used. Integration requires the vector to include a fragment of DNA, ideally no shorter than ~400 bp (although sequences as short as 150 bp have been used), that is homologous to DNA on the *B. subtilis* chromosome. In many cases, an inducible promoter (e.g. P_{spac} or P_{xyl}) is located upstream of the cloning site to control the expression downstream genes in the integrant (see Figure 4.1). The integration vector must encode antibiotic-resistance genes that function in *E. coli* and *B. subtilis*; in the latter case, genes encoding chloramphenicol, kanamycin/neomycin, erythromycin, spectinomycin and tetracycline resistance are available for this purpose. After propagation of the vector in *E. coli*, competent *B. subtilis* cells are transformed with the construct. Because the vector is unable to replicate in *B. subtilis*, selecting for the cognate antibiotic resistance identifies transformants in which the entire plasmid has integrated into the chromosome as a result of a single-crossover recombination between the homologous DNA sequences on the vector and the chromosome. The resulting integrant contains an entire copy of the vector and a duplication of the vector DNA (Figure 4.1). Integrants are relatively stable, with an excision frequency $>10^{-5}$ per cell generation.

Following integration, the functionality of the target gene is dependent on the structure of the cloned fragment. If the fragment carries an intact gene, two functional copies of the target genes will be present in the integrant (Figure 4.1A). The upstream copy is transcribed from its native promoter, the downstream copy by the promoter on the vector. If the fragment contains one or other of the ends of target gene, then one functional and one partially deleted copy will be present on the chromosome (Figure 4.1B and C). The location of the active gene (upstream or downstream) and its promoter is determined by the location of the fragment. However, if the fragment contains sequences that are internal to the target gene, no functional copies will be present after integration (Figure 4.1D).

Single-crossover recombination was used to generate a series of isogenic mutants in a systematic programme designed to identify the function of genes of unknown function, the so-called 'y' genes. The *Bacillus* functional analysis (BFA) mutant collection was constructed using a common technological platform based around the pMUTIN series of integration/reporter gene vectors (Vagner, Dervyn, & Ehrlich, 1998). pMUTIN vectors have the following properties (see Figure 4.1):

- A ColE1 replication origin that is functional in *E. coli*
- An ampicillin-resistance gene (bla) that is functional in *E. coli* and an erythromycin-resistance gene (erm) that is functional in *B. subtilis*
- A lacZ reporter gene preceded by a functional *B. subtilis* RBS
- An IPTG-inducible P_{spac} promoter (Yansura & Henner, 1984) together with constitutively expressed lacI gene

DNA fragments containing either an internal region of the target gene (knockout mutant—see Figure 4.1D) or the RBS and the 5' end (RBS-fusion mutant—see Figure 4.1B) were cloned into pMUTIN, using *E. coli* as an intermediate host, and transformed into *B. subtilis*, selecting for erythromycin resistance. The integration places the *lacZ* gene within the transcriptional unit of the target gene (Figure 4.1), enabling it to be used to monitor the target gene's promoter. If the target gene has genes that are downstream and in the same transcriptional unit (operon), the P_{spac} promoter facilitates their expression (Figure 4.1). This reduces the impact of potential polar effects and allows pMUTIN to be used even when genes downstream are essential for growth. If the target gene itself is essential, as determined by an inability to isolate integrants, its expression can be made conditional by generating an RBS fusion rather than a knockout mutant. In which case, the presence of IPTG is used to control the expression of the intact downstream copy of the target gene (Figure 4.1B).

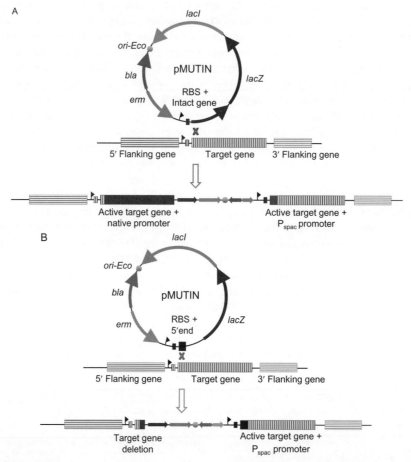

FIGURE 4.1

Integration vectors: integration by single-crossover recombination. A pMUTIN-type vector is used to illustrate the consequences of using different fragments of a target gene on the resulting integrant. Plasmid sequences are shown in solid colour and chromosomal sequences in hatched colours. Promoters (native in colour and vector [P_{spac}] in black) are shown as arrowheads. (A) Integration vector with a complete copy of the target gene, showing two active copies of the target in the integrant, controlled by the native promoter the other by the vector promoter. (B) Integration vector with the native RBS and 5′ end of the target gene, showing one active copy of the target genes controlled by the vector promoter.

(Continued)

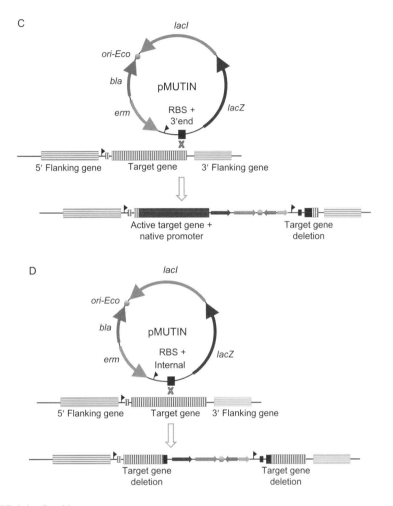

FIGURE 4.1—Cont'd

(C) Integration vector with the 3' end of the target gene, showing one active copy of the target genes controlled by the native promoter. (D) Integration vector with an internal fragment of the target gene, showing inactive copies of the target genes. In all cases, the vector promoter can control the expression of genes downstream of the target gene. (See color plate.)

The BFA mutant collection consisting of over 2500 mutants (Kobayashi et al., 2003) is available from the Japanese National BioResource Project (www.shigen.nig.ac.jp/bsub). The collection was used to identify the phenotypes of many 'y' genes (Biaudet-Brunaud et al., 2001a, 2001b; Sonenshein, Hoch, & Losick, 2002) and to identify 271 genes that were essential for growth on a complex medium. However, the existence of paralogues, and the limited range of conditions that were applied, restricted the extent to which phenotypes and/or functions could be ascribed to particular genes and alternative approaches (both bioinformatical and experimental) have been applied subsequently. The most up-to-date list of gene functions is at SubtiWiki website (http://www.subtiwiki.uni-goettingen.de; Mäder, Schmeisky, Flórez, & Stülke, 2012).

A specialised use of single-crossover recombination is the generation of 'clean' deletion mutations (Figure 4.2) in which target sequences are removed from the

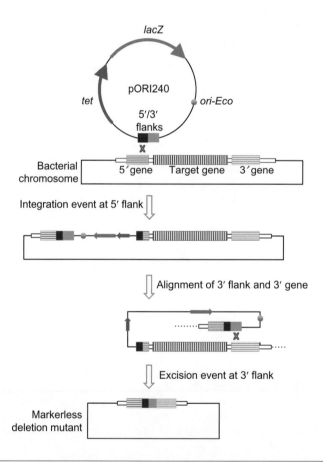

FIGURE 4.2

Integration vectors: Generation of scarless markerless deletions. The pORI1240 vector is used to generate markerless deletion mutations. DNA sequences flanking the target genes or sequences are cloned onto pORI1240. An integration event at one of these flanking sequences (here the 5′ flank) integrates the vector into the chromosome. In the absence of section for the tetracycline resistance, intramolecular excision events can occur, either reversing the original integration event or not involving these flanks (as shown here), generating a scarless, markerless mutation. (See color plate.)

chromosome without their replacement with a marker gene (Leenhouts et al., 1996). One such system is the pORI series of vectors, based on a derivative of pWV01 that lacks *repA*, the gene encoding the replication initiation protein. These vectors can therefore only replicate in helper strains that provide RepA in *trans*. One such integration vector, pORI240 (Figure 4.2), encodes a tetracycline-resistance gene (tet), a *lacZ* reporter gene (encoding β-galactosidase) capable of expression in *B. subtilis* and a multiple cloning site (MCS). Sequences flanking the gene to be deleted are cloned together into the MCS. After amplification in a helper strain providing the RepA function, the vector is transformed into *B. subtilis* and integrants resulting from a single-crossover recombination selected on tetracycline plates. Figure 4.2 shows a crossover between the sequences at the 5' flanking ends of the target gene; however, the crossover could have occurred with equal frequency between sequences at the 3' flanking ends. The transformants are then checked for the production of β-galactosidase *via* the production of blue colonies on agar plates containing the chromogenic substrate X-gal (5-bromo-4-chloro-3-indoylgalactoside). The final step is to screen for colonies resulting from an excision event between the flanking ends *not involved* in the original crossover recombination, the 3' flanking ends in the case of Figure 4.2. Since this occurs at a low frequency and cannot be selected directly, the excision event is detected by screening for white colonies on X-gal-containing plates, indicating that the plasmid sequences have been deleted. The colonies should also be tetracycline-sensitive. Two types of excision event are possible; one is simply a reversal of the original integration event, leaving the target gene intact, and the other, involving the flanking regions not involved in the original crossover recombination, will contain the markerless deletion. These two event types can be distinguished either by loss of the target gene function or by a diagnostic PCR across the flanking sequences.

In the case of double-crossover recombination, the integration vector is used to replace a region of the chromosomal target with either heterologous DNA or modified homologous DNA. Sequences (≥ 400 bp) flanking the chromosomal target are amplified and incorporated into the integration vector. The DNA that is to replace the target DNA, together with a gene encoding a suitable selectable marker, is then cloned between the flanking sequences. Following amplification of the vector in *E. coli*, and transformation into *B. subtilis*, a double-crossover recombination between the vector and target sequence on the chromosome leads to the integration of the sequences between the flanking regions (Figure 4.3). Linearising the vector before transformation forces the selection of integrants that have resulted from a double-rather than single-crossover recombination, since the latter would be lethal to the host. Alternatively, the flanking regions and inserts can be generated by a fusion PCR reaction and the resulting amplicon used for the transformation. In this case, the flanking regions should be at least 1000 bp in length to improve the recombination frequency.

When cloning heterologous DNA into the *B. subtilis* chromosome, the *amyE* locus, encoding an α-amylase, is often used as the target insertion site. Consequently, there are various insertion vectors (e.g. pDR111) (Ben-Yehuda, Rudner, & Losick,

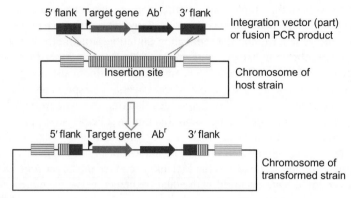

FIGURE 4.3

Integration vectors: integration by double-crossover recombination. The integration vector has 5′ and 3′ flanks that are homologous to the integration site. Between these flanks are a target gene and, if required, a promoter (arrowhead) and an antibiotic-resistance gene (Ab^r). A reciprocal double-crossover recombination removes sequences at the integration site and replaces them with the sequences on the vector. If *amyE* is used as the integration site, the absence of α-amylase can be used to confirm the integration event. (See color plate.)

2003) in which fragments encoding of the front (5′) and back (3′) ends of the *amyE* gene are already incorporated. Between the flanking *amyE* sequences are a selectable antibiotic-resistance gene and an inducible promoter. Integration places the cloned fragment, which may be native or foreign DNA, within the *amyE* gene. An advantage of using the *amyE* locus is that the integration event leads to the inactivation of *amyE* and this provides a useful phenotype that can be screened on starch plates stained with iodine; wild-type cells will exhibit a zone of starch hydrolysis around the colonies, and those with the integrated fragment will not. The method is used for the stable introduction of a single copy of native and foreign DNA, for instance, for gene expression or complementation studies.

Double-crossover recombination can also be used to create markerless mutants, and in recent years, a number of systems have been developed, based on site-specific recombination systems. Two such systems are the Cre/*lox* (Yan, Yu, Hong, & Li, 2008) and RipX/CodV (Xer-cise) (Bloor & Cranenburgh, 2006) systems. Both require the selective markers used to identify the initial insertion event to be flanked by recognition sites for their cognate site-specific recombinases, namely, Cre or RipX/CodV. Following recombination, both systems leave a small 'scar' of additional DNA, 34 bp in the case of the Cre/*lox* system and 28 bp in the case of the Xer-cise system. In both cases, the transforming DNA can be cloned into purpose-designed vectors or simply generated by a splicing/joining PCR.

The Cre/*lox* system involves the highly efficient site-specific Cre recombinase derived from the coliphage P1 (Abremski, Hoess, & Sternberg, 1983). Cre catalyses a reciprocal site-specific recombination between two *loxP* sites and does so independently of any host cofactors or accessory proteins. Therefore, a cassette consisting of an antibiotic-resistance gene (e.g. spectinomycin resistance, *spc*) flanked by *loxP* sequences is placed between DNA sequences (≥400 bp) that flank the target gene(s) to be deleted (Figure 4.4). Following the selection of transformants in which the cassette has replaced the target gene(s) *via* a double-crossover

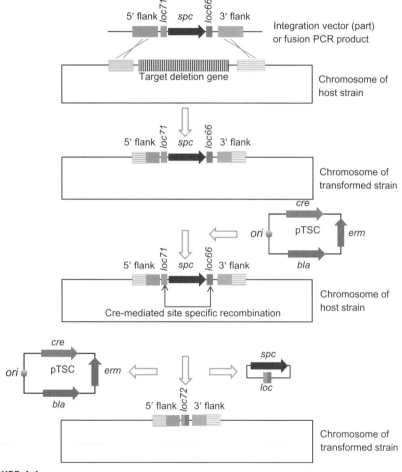

FIGURE 4.4

Integration vectors: Generation of scarred markerless deletions. A cassette including 5′ and 3′ flanks of the target gene, mutant *lox* sites and a selective maker is transformed into the host, either on a linearised plasmid or on fused PCR product. Selection for the antibiotic resistance of the cassette (here spectinomycin resistance or *spc*) gives transformants in which the target gene is deleted. The introduction of the *cre* gene leads to the removal of the antibiotic-resistance marker, leaving behind a double-mutated *lox* site (*lox72*) that is not recognised by Cre. (See color plate.)

recombination, a temperature-sensitive helper plasmid (pTSC) encoding a constitutive *cre* gene is introduced into the strain. The presence of Cre mediates the reciprocal recombination between two *loxP* sites, resulting in the loss of the antibiotic-resistance cassette and leaving a single *loxP* site. In the case of multiple gene deletions, *loxP* sites would accumulate in the strain leading to unwanted recombination events in the presence of Cre. Therefore, Yan and colleagues (2008) used two variants of *loxP*, *lox66* and *lox71*, that had single-base changes in their sequences (Figure 4.4). Although individually these single-base changes do not affect the recombination efficiency of Cre, when recombined during the excision event, the resulting double mutation in the scar sequence, *lox72*, is refractive to Cre-mediated recombination. To reduce the time taken to generate such mutations, the transforming DNA can be added to a cell already containing the helper plasmid. In this case, the *cre* gene needs to be under the control of a tightly controlled inducible promoter.

One disadvantage of the Cre/*lox* system is that the need for a helper plasmid requires additional manipulations. The Xer-cise system uses the native RipX/CodV recombinase that resolves interlocked chromosome dimers (Sciochetti, Piggot, & Blakely, 2001). The target for the RipX/CodV recombinase is a 28 bp *dif* sequence. Similar to the Cre/*lox* system, an antibiotic-resistance cassette is generated with flanking *dif* sites, and this is inserted between DNA sequences (≥ 450 bp) that flank the target gene(s). Once antibiotic-resistance transformants are selected, growth overnight in broth in the absence of the antibiotic results in the spontaneous RopX/CodY-mediated excision of the cassette in a significant proportion of the population ($>20\%$), but leaving behind a potentially active 28 bp *dif* scar.

Although the *Bacillus* functional analysis programme identified 271 genes that were essential for growth in a complex medium, this very much underestimates the number of essential functions encoded by bacterial cells. In the first place, the number of genes that are essential in a chemically defined medium is considerably more as, for example, the cell has to make all of the amino acids required for protein and cell wall synthesis. Secondly, it is now clear that there is significant functional redundancy for key steps in the biosynthesis of many essential cellular components, such as the cell envelope. In order to explore the issue of redundancy in *B. subtilis*, Claessen and colleagues (2008) adapted a genetic screen used to identify synthetic lethal genes in yeast (Bender & Pringle, 1991). To this end, they created the unstable vector pLOSS* (*l*ethal *o*r *s*ynthetic *s*ick) encoding a constitutively expressed β-galactosidase (*lacZ*) gene. A potentially redundant gene is cloned into pLOSS* and the resulting construct transformed in a mutant strain lacking the gene. If the gene is genuinely redundant, continuing to culture the cells will eventually lead to its loss of the pLOSS* construct from the population. The pLOSS*-containing strain can now be used to screen for synthetic lethal or sick strains by identifying genes that compensate for the loss of the potentially redundant gene. The pLOSS* strain is first transformed with a plasmid encoding the mariner transposon (Le Breton, Mohapatra, & Haldenwang, 2006) while maintaining selection for pLOSS*. The cells are allowed to grow for many generations to generate a library of individual transposon-mediated mutation and then screened on nonselective plates containing the X-gal. The vast majority of the colonies will be white because they will not have mutations in a gene that are essential in the absence of the target redundant gene and will have lost the pLOSS* vector. However, blue colonies will have maintained the pLOSS* vector in the absence of selection because of a mutation in a gene capable of compensating for the target redundant gene.

3.6 Expression vectors

A variety of systems have been developed for controlled, high-level expression of genes in *B. subtilis*. Expression vectors can be based on either high-copy-number plasmids or integration plasmids; the latter tend to be more stable in the absence of selection but, having only single copies of the target gene, are generally not expressed to such a high level. The structural stable expression vectors, pHT01 and pHT43, are based on the *E. coli/B. subtilis* shuttle vector pMTLBS72 (Nguyen et al., 2005). pHT01 mediates high-level expression of recombinant proteins into the cytoplasm, while pHT43, which encodes the signal peptide from α-amylase AmyQ, mediates secretion into the culture medium.

Large-scale industrial applications tend to use native promoters capable of directing the sustained synthesis of extracellular proteins during stationary phase and capable of achieving concentrations in excess of 20 g/l. For research purposes and smaller-scale high-added-value fermentations, a range of promoter are available. The two most commonly used are based on the *E. coli* lactose repressor/operator and the *B. subtilis* xylose repressor/operator.

The widely used P_{spac} promoter was generated by fusing the 5' sequences of a promoter from the *B. subtilis* phage SPO1 and the 3' sequences of the *E. coli lac* promoter, including the operator region (Yansura & Henner, 1984). The regulation of P_{spac} requires the presence of the *lacI* gene encoding the *E. coli* lactose repressor, modified to facilitate expression in *B. subtilis*. The level of inducibility can be controlled by providing different concentrations of the inducer IPTG, usually in the range from 0.1 to 10 mM. In a variation of P_{spac} promoter, called $P_{hyperspank}$, changing the G at position −1 (relative to the transcription start site) to a *T* increased the level of expression 10–20-fold without affecting the fold induction by IPTG.

The pHT series of vectors (see earlier) use the IPTG-inducible P_{grac} promoter, consisting of the strong native promoter from the *groESL* operon fused to the *lacO* operator (Phan, Nguyen, & Schumann, 2005).

Xylose-inducible promoters (P_{xyl}) can be used to control gene expression in *B. subtilis* without modification (Gartner, Degenkolb, Ripperger, Allmansberger, & Hillen, 1992). Transcription from the P_{xyl} promoter/operator is controlled by the chromosome-encoded XylR repressor. The addition of xylose inactivates XylR and induces P_{xyl} expression. If P_{xyl}-regulated genes are used on a high-copy-number expression vector, a copy of the *xylR* gene needs to be included on the vector to ensure the promoter is tightly regulated. XylR-controlled promoters direct moderately high levels of expression.

3.7 Genome minimalisation

One widely used approach to strain development is to minimise the genomes of established industrial microorganisms, and just such an approach has been applied to a range of microbes including *B. subtilis*, *Clostridium glutamicum*, *E. coli*, *Schizosaccharomyces pombe* and *Saccharomyces cerevisiae* (Moya et al., 2009). Genome reduction seeks to develop strains with characteristics that are more favourable for industrial processes by addressing the question of what is the minimal set of genes required for the efficient conversion of substrate to product in an industrial bioreactor. While such programmes have been successful in deleting large sections of the *B. subtilis* genome that were perceived to be not essential for growth in conventional culture media (Westers et al., 2003; Manabe et al., 2011), there are few reports concerning the characteristics of these strains in an industrial setting. A combination of systems and synthetic biology provides the opportunity to minimalise the *B. subtilis* genome in a more directed way, by using models to design the minimalisation programme and predict the impact of deletions on growth and metabolism. The use of markerless deletion protocols (see earlier) is essential for such programmes.

4 ANALYSIS OF THE TRANSCRIPTOME
4.1 Transcription and transcription profiling

Transcription in *B. subtilis* is controlled by a combination of sigma factors, regulatory proteins, small untranslated RNA and riboswitches. A systematic and quantitative study, carried out by the European BaSysBio consortium, using high-density tiled arrays (Rasmussen, Nielsen, & Jarmer, 2009), mapped the *B.* subtilis transcriptome under 104 environmental and nutritional conditions (Nicolas et al., 2012). Only 4% of the known coding sequences (CDS) failed to be expressed under any of the conditions used, and these CDS were generally of unknown function and predicted to have originated via horizontal gene transfer. In contrast, only 144 CDS were highly expressed under all of the conditions, indicating that most of the CDS in *B. subtilis* are differentially expressed. This highly expressed group encoded proteins with essential functions and enzymes involved in glycolysis, iron–sulphur metabolism and detoxification. The BaSysBio study identified 32,442 promoters, grouping 2935 of these promoters into regulons controlled by specific sigma factors (Nicolas et al., 2012). The data from these transcriptomic experiments can be downloaded from the BaSysBio website (http://genome.jouy.inra.fr/basysbio/bsubtranscriptome) and viewed at the *B. subtilis* Expression data browser (http://genome.jouy.inra.fr/cgi-bin/seb/index.py). The latter is particularly useful for identifying the transcription initiation and termination sites for genes and groups of genes and potential antisense RNA that may interfere with expression. The transcription profiles can also be viewed from within SubtiWiki (http://www.subtiwiki.uni-goettingen.de), which provides additional information about each gene together with additional references.

4.2 Sigma factors

B. subtilis encodes 17 well-studied sigma factors (Haldenwang, 1995; Helmann, 2002), a PBSX prophage-mediated sigma factor (McDonnell, Wood, Devine, & McConnell, 1994), and a coregulated sigma factor, YvrI-YvrHa, involved in the induction of oxalate decarboxylase in response to acid stress (MacLellan, Guariglia-Oropeza, Gaballa, & Helmann, 2009; MacLellan, Helmann, & Antelmann, 2009). Among the well studied are a group of seven extracytoplasmic function (ECF) sigma factors (Helmann, 2002) that are often cotranscribed with membrane-bound anti-sigma factors that prevent their interaction with RNA polymerase until a cognate signal is received from the environment.

Sigma A (SigA or σ^A, consensus; TTGACA-17 bp-TAATAT), the main vegetative sigma factor responsible for the induction of household genes, shows considerable homology with the vegetative sigma factor across all bacterial species (Iyer & Aravind, 2012). The BaSysBio transcriptional analysis (Nicolas et al., 2012) identified 1859 SigA-controlled promoters. When the sequences associated with these promoters were subjected to an unsupervised hierarchical cluster analysis, six SigA clusters, each with distinct consensus motifs, were identified. The implications of these clusters have still to be understood.

In addition to individual specific stress responses, *B. subtilis* encodes more than 160 general stress proteins that provide nonspecific resistance to stress by protecting DNA, membranes and proteins from the damage (Nannapaneni et al., 2012). The genes encoding these proteins are members of the SigB (σ^B) general stress regulon that includes ATPase-dependent chaperones and associated proteases, enzymes involved in protection from oxidative stress and proteins involved in drug efflux and the uptake of osmoprotectants.

SigA and SigB are two of 10 regular sigma factors. Of the remaining eight, four are sporulation-specific (σ^E, σ^F, σ^G and σ^K): SigH (σ^H) is involved in the regulation of transition-phase activities such as sporulation initiation and competence, SigD (σ^D) controls chemotaxis and motility, SigL (σ^L) is involved in levanase and amino acid catabolism, and SigM (σ^M) is involved in salt resistance.

The remaining seven sigma factors, SigI, SigM, SigV, SigW, SigX, SigY and SigZ, are members of a diverse family of sigma factors, the ECF group. Several of these sigma factors are controlled by cognate anti-sigma factors transcribed from the same operon as the sigma factor itself. In the well-studied SigV and SigW systems, the cognate anti-sigma factors are controlled by 'Regulated *I*ntramembrane Proteolysis', involving the so-called site-1 and site-2 membrane proteases. In response to specific extracellular stresses, the site-1 protease cleaves the extracytoplasmic domain of the anti-sigma factor, rendering it susceptible to intramembrane cleavage by the site-2 protease. This results in the release of the anti-sigma factor/sigma factor complex into the cytoplasm, where the former is further fully degraded. The released sigma factor is then free to initiate the transcription of members of its regulon. For example, SigV (σ^V), responsible for lysozyme resistance, is controlled by the proteolytic destruction of the anti-σ factor RsiV (Hastie, Williams, &

Ellermeier, 2013). The SigW (σ^W) regulon includes >50 genes activated by cell wall stress (Zweers, Nicolas, Wiegert, van Dijl, & Denham, 2012). The *sigW* gene is cotranscribed with its cognate anti-σ factor gene, *rsiW*, from a single transcription start site.

4.3 Transcription termination

Transcription is terminated at transcription termination elements (TTEs). *B. subtilis* uses both intrinsic Rho-independent and Rho-dependent TTEs, but, in contrast to *E. coli*, the Rho factor is not essential for growth in rich media, a fact exploited by the BaSysBio consortium to analyse Rho-dependent terminators. When the whole-genome transcription of a Rho mutant was analysed, Rho-dependent mRNA transcripts were extended by on average 2.5 kb, in some cases up to 12 kb (Nicolas et al., 2012).

4.4 Reporter gene technology

A wide variety of reporter genes have been used to monitor gene expression and protein location in *B. subtilis*. These include *lacZ*, encoding the β-galactosidase from *E. coli*; *bgaB*, encoding the heat-stable β-galactosidase from *Geobacillus stearothermophilus*; *luxAB*, encoding the luciferase from *Vibrio harveyi; mCherry,* encoding an enhanced red fluorescent protein from *Discosoma sp.;* and, by far the most versatile, *gfp,* encoding the green fluorescent protein (GFP) from *Aequorea victoria*. The selection of fluorescent proteins and their variants for imaging studies has been reviewed by Shaner, Steinbach, and Tsien (2005).

4.5 Reporter gene libraries/live cell arrays

The *Bacillus* functional analysis programme used the pMUTIN series of integration vectors (Vagner et al., 1998) to transcriptionally fuse the *lacZ* gene to more than 2000 target genes (Figure 4.1), generating the BFA library of mutants (Kobayashi et al., 2003). The collection was used to determine the level and timing of target expression in minimal and nutrient media by taking samples during growth and carrying out individual β-galactosidase assays on each sample. By comparing the transcription patterns of genes adjacent to each other on the chromosome, putative assignments were made to operons (Biaudet-Brunaud et al., 2001a, 2001b; Prágai, Eschevins, Bron, & Harwood, 2001). The BFA reporter gene library has been used extensively in screening programmes to identify new members of stimulons and regulons, confirming the assignment with the use of null mutations in the cognate regulatory proteins (Prágai & Harwood, 2000).

More recently, the BaSysBio consortium constructed a live cell array to facilitate the high-throughput real-time analysis of gene expression in *B. subtilis* (Botella et al., 2010). The vector used, pBaSysBioII, was an integrative plasmid with a ligation-independent cloning (LIC) site. The use of LIC allowed the cloning process to be automated. The reporter used in this library was a stable variant of the GFP, namely, GFPmut3, with an estimated half-life in *B. subtilis* of ~10 h. The *gfpmut3* gene was preceded by a strong RBS that was complementary to the 3′ end of 16S rRNA. Promoter fragments of at least 400 bp were cloned immediately upstream of the RBS by LIC and the resulting recombinant plasmids integrated into the chromosome by a single-crossover integration across the promoter fragment, an event that was not mutagenic. The resulting array, consisting of more than 800 strains, was used to monitor gene expression in real time using microtitre plates and a Biotek Synergy 2 incubating plate reader. Promoter activity (PA) at a given time point was proportional to the specific GFP signal:

$$PA = \frac{(dGFP/dt)}{OD}$$

where t, time; GFP, measured GFP fluorescence; and OD, culture optical density. As the growth and GFP signals can be noisy, the PA curve can be smoothed using a moving window averaging system by calculating GFP production at $t-1$, t and $t+1$ as follows:

$$PA'(t) = AVERAGE(PA(t-1); PA(t); PA(t+1))$$

4.6 Analysis of populations and single cells

Fluorescent reporters are widely used for studying gene expression and protein locations in populations of cell and in individual cells (see Chapter 6). These analyses can be carried out using flow cytometry, fluorescent microscopy or a combination of both (Yepes et al., 2012; Trip, Veening, Stewart, Errington, & Schaffers, 2013). The range of emission spectra of the available fluorescent proteins (Shaner et al., 2005) means that cells can be dual-labelled to follow the simultaneous expression and/or location of more than one protein. Flow cytometry facilitates the rapid analysis of a very large number of cells in the population, while time-lapse microscopy allows the fate of individual cells to be monitored over time. Time-lapse images can also be used to monitor dynamic movements of proteins within a single cell (Marston & Errington, 1999; Anderson, Gueiros-Filho, & Erickson, 2004; Johnson, van Horck, & Lewis, 2004; Doubrovinski & Howard, 2005).

5 ANALYSIS OF THE PROTEOME

Proteome analysis provides the clearest information about the cell's protein composition. It is able to catalogue, for any growth of growth phase condition, the proteins associated with the cytoplasm, the membrane and the extracytoplasmic environment. Broadly speaking, two experimental approaches are used, gel-based proteomics and gel-free proteomics. Gel-based proteomics used two-dimensional gel electrophoresis in which individual proteins are separated on the basis of charge (pI) and molecular size. Although reproducibility requires strict adherence to protocols, the technique is relatively straightforward. It has the advantages of being relatively inexpensive and facilitates the direct visualisation of posttranslational modifications, such as proteolytic cleavage, phosphorylation and glycosylation, the latter after staining with specific stains. A major limitation is the difficulty of analysing hydrophobic proteins, such as those found in the membrane. A theoretical two-dimensional protein map has been constructed from the calculated pI and molecular mass values of each of the polypeptides encoded by the genome of *B. subtilis* strain 168 (Bernhardt et al., 2001) and this has been reinforced by extensive practical studies. Proteins may be radiolabelled (^{35}S-methionine) or stained with Coomassie Brilliant Blue or silver. Images can be false-coloured so that direct comparisons can be made with between separate gels. Individual polypeptides can be excised from the gel and trypsin digested and identified by mass spectrometry (MS).

Gel-free proteomics represents a major advance, brought about by developments in MS. While gel-based proteomics starts with intact protein, shotgun gel-free proteomics starts by proteolytically digesting the extracted complex mixture of proteins. Multidimensional liquid chromatography coupled with tandem mass spectrometry (MS/MS) is used to size the resulting peptide fragments. The peptide data are fed into a bioinformatics data processing pipeline to identify the protein composition of the original mixture.

In recent years, it has been possible to obtain absolute quantities of the proteins in a mixture. Various approaches have been developed and these are outlined in an excellent review by Becher and colleagues (Otto, Beernhardt, Hecker, Volker, & Becher, 2012) that also includes a detailed analysis of the various methods.

6 ANALYSIS OF THE METABOLOME

The high commercial importance of *B. subtilis* has led to the development of a wide range of strategies for engineering pathways for this bacterium in an effort to increase the production of high-added-value and industrial-scale products. The availability of the complete genomic sequence for *B. subtilis* 168 paved the way for metabolic reconstructions of the global metabolic network of the organism by providing a predicted complete enzyme complement for the strain. Once the enzymes were identified, the metabolite components were identified by reference to existing databases such as KEGG (Kanehisa & Goto, 2000) or from biochemical analysis in the literature. The availability of metabolic reconstructions then facilitates the use of metabolic modelling to make predictions about the most optimal growth conditions for product formation and for possible genetic modifications strategies, such as gene knockouts or gene overexpression, to maximise the yield of a given desired metabolic product. Two genomic-scale metabolic reconstructions of *B. subtilis* have been published and are commonly used in metabolic modelling projects. Oh, Palsson, Park, Schilling, and Mahadevan (2007) produced a genomic-scale reconstruction of the *B. subtilis* metabolic network using high-throughput phenotyping experiments, together with gene essentiality data, to add missing reactions. *i*Bsu1103 is a genome-scale metabolic model of *B. subtilis* based on SEED annotations, an accurate method of functional annotation, that was developed using experimental datasets from over 1500 conditions and employing an improved model optimisation strategy. *i*Bsu1103 includes 1437 reactions associated with 1103 genes and is available as an SBML (Keating, Bornstein, Finney, & Hucka, 2006) file that can be imported into a range of software such as the COBRA toolbox (Schellenberger et al., 2011) and OptFlux (Rocha et al., 2010). These genomic-scale models, and others, can be used with simulation techniques and algorithms such as elementary mode analysis (Papin, Price, & Palsson, 2004), flux balance analysis (Stelling, Klamt, Bettenbrock, Schuster, & Gilles, 2002) and dynamic modelling (Buescher et al., 2012).

7 PARTS, DEVICES, SYSTEMS AND APPLICATIONS
7.1 Parts, systems and devices

iGEM is the International Genetically Engineered Machines competition (http://igem.org/Main_Page). Every year, student teams from around the world compete to design useful applications using synthetic biology approaches. While the majority of the teams select *E. coli* as their chassis, over the years, *B. subtilis* is becoming increasingly popular as a chassis for a range of projects. The BioBrick parts, devices and systems that arise from iGEM and participating laboratories are deposited in a large repository, the Registry of Standard Biological Parts, which provides these parts for use by the community (http://parts.igem.org/Main_Page).

The Cambridge team first proposed the use of *B. subtilis* as a chassis, in the 2007 iGEM competition. The following year saw the use of *B. subtilis* by a number of teams including Newcastle, Cambridge and Imperial College. By 2009, the number of teams using this organism had grown considerably with more than 10 teams developing parts and applications using *B. subtilis* or other *Bacillus* strains (http://2009.igem.org/Main_Page). In 2010, the Imperial College team reached the finals of the competition using *B. subtilis* as a reporter to detect the *Schistosoma* parasite (http://2010.igem.org/Team:Imperial_College_London/Schistosoma) and the Newcastle University team gained a gold medal for their project, BacillaFilla, using *B. subtilis* to repair cracks in concrete (http://2010.igem.org/Team:Newcastle). In 2012, the Munich team worked on a project to enhance the amount *B. subtilis* resources in the registry that included the development of a BacillusBioBrickBox with a range of extra parts (http://2012.igem.org/Team:LMU-Munich). Team Groningen also won the World Championship Jamboree iGEM competition in 2012 using *B. subtilis* to produce a biosensor that would detect spoilage in meat (http://2012.igem.org/Team:Groningen).

As a result of the iGEM team's activity, there are a variety of parts available for *B. subtilis* in the Registry of Standard Biological Parts. Despite the popularity of this organism, an initial chassis-based search of the registry (as of July 2013) produces a very limited range of parts: 13 promoters (SigA and SigB regulated), 6 RBSs and 23 coding sequences of various functions. A more detailed search reveals that there are indeed many more parts, but they are not annotated specifically for use in *B. subtilis*, and a free text-based search for *B. subtilis* parts reveals 319 parts with '*Bacillus subtilis*' in their textual description. Indeed, a significant number of the finalist iGEM projects from 2008 to 2013 have been based on *B. subtilis*. While the Registry of Standard Biological Parts provides a huge range of potential resources, many of the parts listed still remain to be characterised experimental and their function verified.

The Flowers Consortium (www.synbiuk.org) aims to produce the UK infrastructure for Synthetic Biology. *B. subtilis* is one of the chosen host organisms on which the project will focus and therefore, in the future, we can look forward to an increase in the number and range of parts and systems available for this organism (http://intbio.ncl.ac.uk/?projects=an-infrastructure-for-platform-technology-in-synthetic-biology).

7.2 Engineering genomes with *B. subtilis*

B. subtilis is an attractive chassis for the assembly of genome-scale heterologous DNA fragments. For example, Itaya, Tsuge, Koizumi, and Fujita (2005) demonstrated the stable cloning of the genome of *Synechocystis* PCC6803 in the genome of *B. subtilis* 168. More recently, Iwata and colleagues (Watanabe, Shiwa, Itaya, & Yoshikawa, 2012) have developed a novel cloning system for large DNA fragments for engineering genome-scale segment in *B. subtilis*. In their *B. subtilis* genome (BGM) vector, the entire 4.2 Mb genome of *B. subtilis* functions as a vector. The BGM vector system allows for a cloning capacity of over 3 Mb, stable propagation of cloned DNA and was used to clone a 158 gene segment of the fishlike odorant receptor (class I OR) gene family, which was then shown to generate transgenic mice.

7.3 Biosensors

B. subtilis has tremendous potential as a chassis for synthetic biosensors and biosensors are currently one of the most common synthetic biology applications developed for this organism. Not only does the organism possess a wide range of natural two-component systems and quorum-sensing systems that can be modified, but also its ability to sporulate provides a mechanism for the storage and shipment of systems as spores. This property allows engineered organisms to be distributed as spores that have a high resilience to temperature and desiccation. These properties are ideal for biosensors that need to be deployed in rural settings in developing countries. For example, French and coworkers (2012) have used *B. subtilis* as a chassis for the development of an arsenic biosensor. In a proof-of-principle exercise, they constructed a system, designated a 'Bacillosensor', that employs the *xylE* reporter gene from *Pseudomonas putida* (encoding catechol-2,3-dioxygenase). The *xylE* gene was placed under the control of the repressible promoter P_{ars}, which is controlled by the arsenic-sensitive repressor, ArsR. The result was a strain that produces a yellow coloration in the medium when exposed to arsenic. This strain is designed to detect the contaminating drinking water in rural environments with arsenic. The presence of the strain as spores means that the sensor can be boiled before use to remove contaminating bacteria. The system was found to detect arsenic levels below the WHO-recommended limit of 10 ppb.

8 COMPUTATIONAL TOOLS AND RESOURCES

Since synthetic biology is still in its infancy, the availability of tools and resources specific for the development of synthetic genetic systems in *B. subtilis* is still quite limited. However, synthetic biology approaches depend heavily on conventional bioinformatic tools. Indeed, there is now a wide range of generic bioinformatic and synthetic biology tools for bacteria, most of which include detailed information about, or are relevant to, *B. subtilis*. However, the availability of the organism's genome sequence and recent developments *in B. subtilis* systems biology have resulted in the production of a number of additional computational tools and resources that are dedicated to *B. subtilis*, and some of the most widely used are discussed in the succeeding text.

Firstly, information about the *B. subtilis* genome sequence is available from a number of sources. Perhaps the most widely known resource, and the reference database, is the SubtiList database (Moszer, Glaser, & Danchin, 1995), which is part of the GenoList family (http://genodb.pasteur.fr/cgi-bin/WebObjects/GenoList). In addition, the BacilluScope database was set up to provide a convenient method of searching the more recent reannotations of the *B. subtilis* genome (Barbe et al., 2009).

In recent years, an emphasis has been placed on community-driven annotation of *B. subtilis* and the SubtiWiki (http://subtiwiki.uni-goettingen.de/) and SubtiPathways (http://subtiwiki.uni-goettingen.de/wiki/index.php/SubtiPathways) resources have been developed that allow an integrated view of *B. subtilis* genes and pathways and allow researchers to curate database entries to keep the information up to date (Lammers et al., 2010; Mäder et al., 2012). Transcriptomics resources from the recent *B. subtilis* systems biology projects, such as BaSysBio, are discussed in the preceding text (see Section 4.1). In addition to these newer resources, DBTBS is a database of regulatory motifs and pathways dedicated to *B. subtilis*. While the database has not been updated since 2007, it still represents a highly useful resource for *B. subtilis* synthetic biology (Sierro, Makita, de Hoon, & Nakai, 2008). For example, BacillusRegNet integrates information from DBTBS with comparative genomics information to make predictions about regulatory networks in closely related species (http://intbio.ncl.ac.uk/?projects=bacillusregnet). This system is very useful for identifying transcriptional factors from strains that are closely related to *B. subtilis* (and so are more likely to function) but that are not orthologous to known *B. subtilis* transcription factors and are less likely to interact with native systems. These transcription factors provide a set of parts for the engineering of orthogonal regulatory control in *B. subtilis*. Similarly, it is also possible to identify transcription factors from *B. subtilis* that can be used to engineer gene regulation in other *Bacillus* species

(Misirli, 2011). Another integrative system that provides a useful source of parts for *B. subtilis* is the BacillOndex system (http://intbio.ncl.ac.uk/?projects=bacillondex). BacillOndex is an integrated knowledge base combining genome annotations with data about genetic regulatory networks, biochemical reactions, microarray experiments and protein–protein interactions (Misirli et al., 2013). The system is built as a semantically enriched network using the Ondex system (Köhler et al., 2006) in which nodes represent biological concepts such as genes and gene products, and the edges represent the relationships between these concepts. Information (including the DNA sequence) about basic genetic parts such as coding sequences, promoters, operators, terminators, RBS and spacer sequences can be extracted for use in *B. subtilis* synthetic biology projects. Furthermore, information about the interactions of gene products and annotations, such as those that are specified by GO terms, can be used to classify biological parts in order to rationally design synthetic genetic circuits.

The computational modelling of synthetic systems prior to their implementation is at the heart of synthetic biology. Many of the tools developed for the systems biology of *B. subtilis* are therefore very useful in this respect. These systems range from pathway modelling tools through to tools for modelling multicellular systems such as biofilms. The industrial importance of *B. subtilis* has resulted in the development of a number of approaches and resources for the engineering of its metabolism, including metabolic modelling, and these are discussed in the preceding text (see Section 6). More generally, tools for the modelling of synthetic systems for *B. subtilis*, including regulatory networks, are starting to become available. Dynamic modelling is an important method for the design of regulatory systems in *B. subtilis*. Hallinan and coworkers have described an approach that uses evolutionary computing and stochastic modelling to engineer the regulatory mechanisms controlling differentiation in *B. subtilis* (Hallinan, Misirli, & Wipat, 2010). The BacilloBricks system provides a repository of dynamic models of *B. subtilis* parts, termed standard virtual parts, that can be composed to produce model-based designs for a desired synthetic system (http://bacillobricks.co.uk). These parts have recently been used for the design of *B. subtilis* two-component system-based biosensors (Hallinan, Gilfellon, Misirli, Phillips, & Wipat, 2013). Population-level modelling is also important for synthetic biology design, especially if the differentiation properties and resulting heterogeneity of the organism are to be either exploited, catered for or removed through engineering approaches. In a recent study (Stiegelmeyer & Giddings, 2013), an agent-based model of the competence phenotype switching was produced. The model predicted three potential sources of genetic noise based on the spatial arrangement of molecules, epigenetic factors based on the concentration of key competence molecules and the dilution of protein concentrations by cell division. This model may also prove useful for engineering at a population level. Synthetic biofilms are an attractive target for engineered biological systems, since they can be used to produce chemicals and biofuels and also implement biological computers. An approach for modelling synthetic biofilms has been described by Rudge, Steiner, Phillips, and Haseloff (2012) in which an extension to the CellModeller tool has been developed to allow designs for synthetic biofilms to be simulated and tested. This tool combines three-dimensional biophysical models of individual cells with models of genetic regulation and intercellular signalling.

SUMMARY

The genetics, biochemistry and physiology of *B. subtilis* has been extensively studied for over 60 years, initially because it was a nonpathogenic bacterium in which to study spore formation, an issue of importance to the food and canning industries. The discovery that *B. subtilis* was genetically transformable, and therefore genetically amenable, meant that is was an ideal model organism in which to study processes that were unique to Gram-positive bacteria. These studies were considerably enhanced by the formation of research communities in Europe and Japan that lead to genome sequencing and systematic functional analysis programmes. Together with protein secretion and, more recently, systematic systems biology programmes, these intensive research efforts have resulted in *B. subtilis* being one of the most important model microbial organisms, along with *E. coli* and *S. cerevisiae*. *B. subtilis* is easy to cultivate and is very amenable to genome engineering and the extensive knowledge of its metabolic and regulatory pathways means that it can be subjected to sophisticated modelling.

Together with its close relative, particularly *B. licheniformis*, *B. subtilis* is used extensively in industry for the production of industrial enzymes, peptide antibiotics and specialist metabolites. The application of synthetic biology approaches to these processes has the potential to increase both the yield and range of products that can be produced from these bacteria.

Strains of another close relative, *B. amyloliquefaciens*, have been developed as a plant probiotic to provide an environmentally sound means of protecting seedling from damping-off fungal infections that cause root rot. The application of synthetic biology to these organisms has the potential to increase crop yield and to provide sensor that detect the presence of chemical and even other organisms in the environment of plants.

We are currently at the early stages of developing *B. subtilis* as a highly adaptable chassis for synthetic biology; however, the potential rewards are likely to be very significant.

References

Abremski, K., Hoess, R., & Sternberg, N. (1983). Studies on the properties of P1 site-specific recombination: Evidence for topologically unlinked products following recombination. *Cell, 32*, 1301–1311.

Anderson, D. E., Gueiros-Filho, F. J., & Erickson, H. P. (2004). Assembly dynamics of FtsZ rings in *Bacillus subtilis* and *Escherichia coli* and effects of FtsZ-regulating proteins. *Journal of Bacteriology, 186*, 5775–5781.

Barbe, V., Cruveiller, S., Kunst, F., Lenoble, P., Meurice, G., Sekowska, A., et al. (2009). From a consortium sequence to a unified sequence: The *Bacillus subtilis* 168 reference genome a decade later. *Microbiology, 155*, 1758–1775.

Bender, A., & Pringle, J. R. (1991). Use of a screen for synthetic lethal and multicopy suppresses mutants to identify two new genes involved in morphogenesis in *Saccharomyces cerevisiae*. *Molecular and Cellular Biology, 11*, 1295–1305.

Ben-Yehuda, S., Rudner, D. Z., & Losick, R. (2003). RacA, a bacterial protein that anchors chromosomes to the cell poles. *Science, 299*, 532–536.

Bernhardt, J., Buttner, K., Coppee, J. Y., Lelong, C., Ogasawara, N., Scharf, C., et al. (2001). The contribution of the European Community consortium to the two-dimensional protein index of Bacillus subtilis. In W. Schumann, S. D. Ehrlich, & N. Ogasawara (Eds.), *Functional analysis of bacterial gene spp* (pp. 63–74). Chichester: John Wiley and Sons.

Biaudet-Brunaud, V., Samson, F., Gas, S., Dervyn, E., Gallezot, G., Duchet, S., et al. (2001a). Phenotype responses and reporter gene activity from the systematic functional analysis of *Bacillus subtilis* unknown genes. In W. Schumann, S. D. Ehrlich, & N. Ogasawara (Eds.), *Functional analysis of bacterial genes* (pp. 53–61). Chichester: John Wiley and Sons.

Biaudet-Brunaud, V., Samson, F., Gas, S., Dervyn, E., Gallezot, G., Duchet, S., et al. (2001b). List of *Bacillus subtilis* genes with a phenotype determined by the functional analysis project. In W. Schumann, S. D. Ehrlich, & N. Ogasawara (Eds.), *Functional analysis of bacterial genes* (pp. 283–292). Chichester: John Wiley and Sons.

Bloor, A. E., & Cranenburgh, R. M. (2006). An efficient method of selectable marker gene excision by Xer recombination for gene replacement in bacterial chromosomes. *Applied and Environmental Microbiology, 72*, 2520–2525.

Botella, E., Fogg, M., Jules, M., Piersma, S., Doherty, G., Hansen, A., et al. (2010). pBaSysBioII: An integrative plasmid generating *gfp* transcriptional fusions for high-throughput analysis of gene expression in *Bacillus subtilis*. *Microbiology, 156*, 1600–1608.

Bron, S. (1990). Plasmids. In C. R. Harwood & S. M. Cutting (Eds.), *Molecular biological methods for Bacillus* (pp. 75–174). Chichester: John Wiley and Sons.

Bruand, C., Ehrlich, S. D., & Janniere, L. (1991). Unidirectional theta replication of the stable *Enterococcus faecalis* plasmid pAM81. *EMBO Journal, 10*, 2171–2177.

Buescher, J. M., Liebermeister, W., Jules, M., Uhr, M., Muntel, J., Botella, E., et al. (2012). Global network reorganization during dynamic adaptations of *Bacillus subtilis* metabolism. *Science, 335*, 1099–1103.

Chang, S., & Cohen, S. N. (1979). High frequency transformation of *Bacillus subtilis* protoplasts by plasmid DNA. *Molecular and General Genetics, 168*, 111–115.

Claessen, D., Emmins, R., Hamoen, L. W., Daniel, R. A., Errington, J., & Edwards, D. H. (2008). Control of the cell elongation–division cycle by shuttling of PBP1 protein in *Bacillus subtilis*. *Molecular Microbiology, 68*, 1029–1042.

Claverys, J. P., Martin, B., & Polard, P. (2009). The genetic transformation machinery: Composition, localization, and mechanism. *FEMS Microbiology Reviews, 33*, 643–656.

Doubrovinski, K., & Howard, M. (2005). Stochastic model for Soj relocation dynamics in *Bacillus subtilis*. *Proceedings of the National Academy of Sciences of the United States of America, 102*, 9808–9813.

Dubnau, D. (1993). Genetic exchange and homologous recombination. In A. L. Sonenshein, J. A. Hoch, & R. Losick (Eds.), *Bacillus subtilis and other gram-positive* bacteria: Biochemistry, physiology, and molecular genetics (pp. 555–584). Washington, DC: American Society for Microbiology.

Dubnau, D., & Lovett, C. M. (2002). Transformation and recombination. In A. L. Sonenshein, J. A. Hoch, & R. Losick (Eds.), *Bacillus subtilis and its closest relatives: From genes to cells* (pp. 453–471). Washington, DC: American Society for Microbiology.

Ehrlich, S. D. (1977). Replication and expression from *Staphylococcus aureus* in *Bacillus subtilis*. *Proceedings of the National Academy of Sciences of the United States of America, 74*, 1680–1682.

Errington, J. (1984). Efficient *Bacillus subtilis* cloning system using bacteriophage vector φ105J9. *Microbiology, 130*, 2615–2628.

Errington, J., & Jones, D. (1987). Cloning in *Bacillus subtilis* by transfection with bacteriophage vector φ105J27: Isolation and preliminary characterization of transducing phages for 23 sporulation loci. *Microbiology, 133*, 493–502.

French, C. E., de Mora, K., Joshi, N., Elfick, A., Haseloff, J., & Ajioka, J. (2012). Synthetic biology and the art of biosensor design. In E. R. Choffnes, D. A. Relman, & L. Pray (Eds.), *The science and applications of synthetic and systems biology: Workshop summary* (pp. 178–201). Washington, DC: The National Academies Press.

Gartner, D., Degenkolb, J., Ripperger, J. A. E., Allmansberger, R., & Hillen, W. (1992). Regulation of the *Bacillus subtilis* W23 xylose utilization operon: Interaction of the Xyl repressor with the *xyl* operator and the inducer xylose. *Molecular and General Genetics, 232*, 415–422.

Gonzalez-Pastor, J. E., Hobbs, E. C., & Losick, R. (2007). Cannibalism by sporulating bacteria. *Science, 301*, 510–513.

Gryczan, T. J., & Dubnau, D. (1978). Construction and properties of chimeric plasmids in *Bacillus subtilis*. *Proceedings of the National Academy of Sciences of the United States of America, 75*, 1428–1432.

Haldenwang, W. G. (1995). The sigma factors of Bacillus subtilis. *Microbiological Reviews, 59*, 1–30.

Hallinan, J. S., Gilfellon, O., Misirli, G., Phillips, A., & Wipat, A. (2013). Tuning receiver characteristics in bacterial quorum communication: An evolutionary approach using standard virtual biological parts. *ACS Synthetic Biology*, in press.

Hallinan, J. S., Misirli, G., & Wipat, A. (2010). Evolutionary computation for the design of a stochastic switch for synthetic genetic circuits. *Conference Proceedings of the IEEE Engineering in Medicine and Biology Society, 2010*, 768–774.

Harwood, C. R., & Archibald, A. R. (1990). Growth, maintenance and general techniques. In C. R. Harwood & S. M. Cutting (Eds.), *Molecular biological methods for Bacillus* (pp. 1–26). Chichester: John Wiley and Sons.

Hastie, J. L., Williams, K. B., & Ellermeier, C. D. (2013). The Activity of σV, an extracytoplasmic function σ factor of *Bacillus subtilis*, is controlled by regulated proteolysis of the anti-σ factor RsiV. *Journal of Bacteriology, 195*, 3135–3144.

Helmann, J. D. (2002). The extracytoplasmic function (ECF) sigma factors. *Advances in Microbial Physiology, 46*, 47–110.

Howell, A., Dubrac, S., Andersen, K. K., Noone, D., Fert, J., Msadek, T., et al. (2003). Genes controlled by the essential YycG/YycF two-component system of *Bacillus subtilis* revealed through a novel hybrid regulator approach. *Molecular Microbiology, 49*, 1639–1655.

Ishii, T., Ki, Yoshida, Terai, G., Fujita, Y., & Nakai, K. (2001). DBTBS: A database of *Bacillus subtilis* promoters and transcription factors. *Nucleic Acids Research*, *29*, 278–280.

Itaya, M., Tsuge, K., Koizumi, M., & Fujita, K. (2005). Combining two genomes in one cell: Stable cloning of the *Synechocystis* PCC6803 genome in the *Bacillus subtilis* 168 genome. *Proceedings of the National Academy of Sciences of the United States of America*, *102*, 15971–15976.

Iyer, L. M., & Aravind, L. (2012). Insights from the architecture of the bacterial transcription apparatus. *Journal of Structural Biology*, *179*, 299–319.

Janniere, L., Bruand, C., & Ehrlich, S. D. (1990). Structurally stable *Bacillus subtilis* cloning vectors. *Gene*, *87*, 53–59.

Janniere, L., Gruss, A., & Ehrlich, S. D. (1993). Plasmids. In A. L. Sonenshein, J. A. Hoch, & R. Losick (Eds.), *Bacillus subtilis and other Gram-positive bacteria: Biochemistry, physiology, and molecular genetics* (pp. 625–644). Washington, DC: American Society for Microbiology.

Johnson, A. S., van Horck, S., & Lewis, P. J. (2004). Dynamic localization of membrane proteins in *Bacillus subtilis*. *Microbiology*, *150*, 2815–2824.

Kanehisa, M., & Goto, S. (2000). KEGG: Kyoto encyclopedia of genes and genomes. *Nucleic Acids Research*, *28*, 27–30.

Kearns, D. B., & Losick, R. (2003). Swarming motility in undomesticated *Bacillus subtilis*. *Molecular Microbiology*, *49*, 581–590.

Keating, S. M., Bornstein, B. J., Finney, A., & Hucka, M. (2006). SBMLToolbox: An SBML toolbox for MATLAB users. *Bioinformatics*, *22*, 1275–1277.

Kobayashi, K., Ehrlich, S. D., Albertini, A., Amati, G., Andersen, K. K., Arnaud, M., et al. (2003). Essential Bacillus subtilis genes. *Proceedings of the National Academy of Sciences of the United States of America*, *100*, 4678–4683.

Koehler, T. M., & Thorne, C. B. (1987). *Bacillus subtilis* (natto) plasmid pLS20 mediates interspecies plasmid transfer. *Journal of Bacteriology*, *169*, 5271–5278.

Köhler, J., Baumbach, J., Taubert, J., Specht, M., Skusa, A., Rüegg, A., et al. (2006). Graph-based analysis and visualization of experimental results with ONDEX. *Bioinformatics*, *22*, 1383–1390.

Kunst, F., Ogasawara, N., Moszer, I., Albertini, A. M., Alloni, G., Azevedo, V., et al. (1997). The complete genome sequence of the Gram-positive bacterium *Bacillus subtilis*. *Nature*, *390*, 249–256.

Lammers, C. R., Flórez, L. A., Schmeisky, A. G., Roppel, S. F., Mäder, U., Hamoen, L. W., et al. (2010). Connecting parts with processes: SubtiWiki and SubtiPathways integrate gene and pathway annotation for *Bacillus subtilis*. *Microbiology*, *156*, 849–859.

Le Breton, Y., Mohapatra, N. P., & Haldenwang, W. G. (2006). In vivo random mutagenesis of *Bacillus subtilis* by use of TnYLB-1, a mariner-based transposon. *Applied and Environmental Microbiology*, *72*, 327–333.

Leenhouts, K. (1995). Integration strategies and vectors. In J. J. Ferretti, M. S. Gilmore, T. R. Klaenhammer, & F. Brown (Eds.), *Genetics of streptococci, enterococci and lactococci* (pp. 523–530). Basel: Karger.

Leenhouts, K. J., Buist, G., Bolhuis, A., ten Berge, A., Kiel, J., Mierau, I., et al. (1996). A general system for generating unlabelled gene replacements in bacterial chromosomes. *Molecular and General Genetics*, *253*, 217–224.

Leenhouts, K. J., & Venema, G. (1993). Lactococcal plasmid vectors. In K. G. Hardy (Ed.), *Plasmids, a practical approach* (pp. 65–94). New York: Oxford University Press.

MacLellan, S. R., Guariglia-Oropeza, V., Gaballa, A., & Helmann, J. D. (2009). A two-subunit bacterial sigma-factor activates transcription in *Bacillus subtilis*. *Proceedings of the National Academy of Sciences of the United States of America, 106*, 21323–21328.

MacLellan, S. R., Helmann, J. D., & Antelmann, H. (2009). The YvrI alternative sigma factor is essential for acid stress induction of oxalate decarboxylase in *Bacillus subtilis*. *Journal of Bacteriology, 191*, 931–939.

Mäder, U., Schmeisky, A. G., Flórez, L. A., & Stülke, J. (2012). SubtiWiki—A comprehensive community resource for the model organism Bacillus subtilis. *Nucleic Acids Research, 40*, D1278–D1287.

Manabe, K., Kageyama, Y., Morimoto, T., Ozawa, T., Sawada, K., Endo, K. T., et al. (2011). Combined effect of improved cell yield and increased specific productivity enhances recombinant enzyme production in genome-reduced *Bacillus subtilis* strain MGB874. *Applied and Environmental Microbiology, 77*, 8370–8381.

Marston, A. L., & Errington, J. (1999). Dynamic movement of the ParA-like Soj protein of *B. subtilis* and its dual role in nucleoid organization and developmental regulation. *Molecular Cell, 4*, 673–682.

McDonnell, G. E., Wood, H., Devine, K. M., & McConnell, D. J. (1994). Genetic control of bacterial suicide: Regulation of the induction of PBSX in *Bacillus subtilis*. *Journal of Bacteriology, 176*, 5820–5830.

Meijer, W. J. J., de Boer, A., van Tongeren, S., Venema, G., & Bron, S. (1995). Characterization of the replication region of the *Bacillus subtilis* pLS20: A novel type of replicon. *Nucleic Acids Research, 23*, 3214–3223.

Misirli, G. (2011). Data integration strategies for informing computational design in synthetic biology. Thesis, Newcastle University.

Misirli, G., Wipat, A., Mullen, J., James, K., Pocock, M., Smith, W., et al. (2013). BacillOndex: An integrated data resource for systems and synthetic biology. *Journal of Integrative Bioinformatics, 10*, 224.

Moszer, I., Glaser, P., & Danchin, A. (1995). SubtiList: A relational database for the *Bacillus subtilis* genome. *Microbiology, 141*, 261–268.

Moya, A., Gil, R., Latorre, A., Pereto, J., Garcillan-Barcia, N. P., & de la Cruz, F. (2009). Toward minimal bacterial cells: Evolution vs. design. *FEMS Microbiology Reviews, 33*, 225–235.

Mozser, I. (2002). *Bacillus subtilis* genome, genes and function. In A. L. Sonenshein, J. A. Hoch, & R. Losick (Eds.), *Bacillus subtilis and its closest relatives: From genes to cells* (pp. 7–11). Washington, DC: American Society for Microbiology.

Nannapaneni, P., Hertwig, F., Depke, M., Hecker, M., Mäder, U., Völker, U., et al. (2012). Defining the structure of the general stress regulon of *Bacillus subtilis* using targeted microarray analysis and random forest classification. *Microbiology, 158*, 696–707.

Nguyen, D. H., Nguyen, Q. A., Ferreira, R. C., Ferreira, L. C. S., Tran, L. T., & Schumann, W. (2005). Construction of plasmid-based expression vectors for *Bacillus subtilis* exhibiting full structural stability. *Plasmid, 54*, 241–248.

Nicolas, P., Mäder, M., Dervyn, E., Rochat, T., Leduc, A., Pigeonneau, N., et al. (2012). Condition-dependent transcriptome reveals high-level regulatory architecture in Bacillus subtilis. *Science, 335*, 1103–1106.

Nijland, R., Burgess, J. G., Errington, J., & Veening, J.-M. (2010). Transformation of Environmental *Bacillus subtilis* Isolates by Transiently Inducing Genetic Competence. *PLoS One, 5*, e9724.

Oh, Y. K., Palsson, B. O., Park, S. M., Schilling, C. H., & Mahadevan, R. (2007). Genome-scale reconstruction of metabolic network in *Bacillus subtilis* based on high-throughput phenotyping and gene essentiality data. *Journal of Biological Chemistry*, *282*, 28791–28799.

Otto, A., Beernhardt, J., Hecker, M., Volker, U., & Becher, D. (2012). *Proteomics: From relative to absolute quantification for systems biology approaches*. Oxford: Elsevier.

Papin, J., Price, N., & Palsson, B. O. (2004). Comparison of network-based pathway analysis methods. *Trends in Biotechnology*, *22*, 400–405.

Phan, T. T. P., Nguyen, H. D., & Schumann, W. (2005). Novel plasmid-based expression vectors for intra- and extracellular production of recombinant proteins in *Bacillus subtilis*. *Protein Expression and Purification*, *46*, 189–195.

Prágai, Z., Eschevins, C., Bron, S., & Harwood, C. R. (2001). *Bacillus subtilis* NhaC, a Na+/H+ antiporter, influences the expression of the *phoPR* operon and the production of alkaline phosphatases. *Journal of Bacteriology*, *184*, 6819–6823.

Prágai, Z., & Harwood, C. R. (2000). Screening mutants affected in their response to phosphate. In W. Schumann, S. D. Ehrlich, & N. Ogasawara (Eds.), *Functional analysis of bacterial genes: A practical manual* (pp. 245–249). Chichester, UK: John Wiley & Sons.

Rasmussen, S., Nielsen, H. B., & Jarmer, H. (2009). The transcriptionally active regions in the genome of *Bacillus subtilis*. *Molecular Microbiology*, *73*, 1043–1057.

Rocha, E. P. C., Danchin, A., & Viari, A. (1999). Analysis of long repeats in bacterial genomes reveals alternative evolutionary mechanisms. *Molecular Biology and Evolution*, *16*, 1219–1230.

Rocha, I., Maia, P., Evangelista, P., Vilaça, P., Soares, S., Pinto, J. P., et al. (2010). OptFlux: An open-source software platform for in silicometabolic engineering. *BMC Systems Biology*, *4*, 45.

Rocha, E., Moszer, I., Klaerr-Blanchard, M., Sekowska, A., Medigue, C., Viari, A., et al. (2001). In silico genome analysis. In W. Schumann, S. D. Ehrlich, & N. Ogasawara (Eds.), *Functional analysis of bacterial genes* (pp. 6–19). Chichester: John Wiley and Sons.

Rudge, T. J., Steiner, P. J., Phillips, A., & Haseloff, J. (2012). Computational modeling of synthetic microbial biofilms. *ACS Synthetic Biology*, *1*, 345–352.

Saier, M. H. J., Goldman, S. R., Maile, R. R., Moreno, M. S., Weyler, W., Yang, N., et al. (2002a). Overall transport capabilities of *Bacillus subtilis*. In A. L. Sonenshein, J. A. Hoch, & R. Losick (Eds.), *Bacillus subtilis and its closest relatives: From genes to cells* (pp. 113–128). Washington, DC: American Society for Microbiology.

Saier, M. H. J., Goldman, S. R., Maile, R. R., Moreno, M. S., Weyler, W., Yang, N., et al. (2002b). Transport capabilities encoded within the *Bacillus subtilis* genome. *Journal of Molecular Microbiology and Biotechnology*, *4*, 37–67.

Schellenberger, J., Que, R., Fleming, R. M. T., Thiele, I., Orth, J. D., Feist, A. M., et al. (2011). Quantitative prediction of cellular metabolism with constraint-based models: The COBRA Toolbox v2.0. *Nature Protocols*, *6*, 1290–1307.

Sciochetti, S. A., Piggot, P. J., & Blakely, G. W. (2001). Identification and characterisation of the *dif* site from *Bacillus subtilis*. *Journal of Bacteriology*, *183*, 1058–1068.

Shaner, N. C., Steinbach, P. A., & Tsien, R. Y. (2005). A guide to choosing fluorescent proteins. *Nature Methods*, *2*, 905–909.

Sierro, N., Makita, Y., de Hoon, M., & Nakai, K. (2008). DBTBS: A database of transcriptional regulation in *Bacillus subtilis* containing upstream intergenic conservation information. *Nucleic Acids Research*, *36*, D93–D96.

References

Sonenshein, A. L., Hoch, J. A., & Losick, R. (2002). Bacillus subtilis *and its closest relatives: From genes to cells.* Washington, DC: American Society for Microbiology.

Spizizen, J. (1958). Transformation of biochemically deficient strains of *Bacillus subtilis* by deoxyribonucleate. *Proceedings of the National Academy of Sciences of the United States of America, 44*, 1072–1078.

Stelling, J., Klamt, S., Bettenbrock, K., Schuster, S., & Gilles, E. D. (2002). Metabolic network structure determines key aspects of functionality and regulation. *Nature, 420*, 190–193.

Stiegelmeyer, S. M., & Giddings, M. C. (2013). Agent-based modeling of competence phenotype switching in *Bacillus subtilis. Theoretical Biology & Medical Modelling, 10*, 23.

Titok, M. A., Chapuis, J., Selezneva, Y. V., Lagodich, A. V., Prokulevich, V. A., Ehrlich, S. D., et al. (2003). *Bacillus subtilis* soil isolates: Plasmid replicon analysis and construction of a new theta-replicating vector. *Plasmid, 49*, 53–62.

Trip, E. N., Veening, J. W., Stewart, E. J., Errington, J., & Schaffers, D.-J. (2013). Balanced transcription of cell division genes in Bacillus subtilis as revealed by single cell analysis. *Environmental Microbiology*, in press. http://dx.doi.org/10.1111/1462-2920.12148.

Vagner, V., Dervyn, E., & Ehrlich, S. D. (1998). A vector for systematic gene inactivation in *Bacillus subtilis. Microbiology, 144*, 3097–3104.

Veening, J. W., Stewart, E. J., Berngruber, T. W., Taddei, F., Kuipers, O. P., & Hamoen, L. W. (2008). Bet-hedging and epigenetic inheritance in bacterial cell development. *Proceedings of the National Academy of Sciences of the United States of America, 105*, 4393–4398.

Verhamme, D. T., Kiley, T. B., & Stanley-Wall, N. R. (2007). DegU co-ordinates multicellular behaviour exhibited by *Bacillus subtilis. Molecular Microbiology, 65*, 554–568.

Vojcic, L., Despotovic, D., Martinez, R., Maurer, K.-H., & Schwaneberg, U. (2012). An efficient transformation method for *Bacillus subtilis* DB104. *Applied Microbiology and Biotechnology, 94*, 487–493.

Watanabe, S., Shiwa, Y., Itaya, M., & Yoshikawa, M. (2012). Complete Sequence of the first chimera genome constructed by cloning the whole genome of *Synechocystis* strain PCC6803 into the *Bacillus subtilis* 168 genome. *Journal of Bacteriology, 194*, 7007.

Westers, H., Dorenbos, R., van Dijl, J. M., Kabel, J., Flanagan, T., Devine, K. M., et al. (2003). Genome engineering reveals large dispensable regions in *Bacillus subtilis. Molecular Biology and Evolution, 20*, 2076–2090.

Wipat, A., & Harwood, C. R. (1999). The *Bacillus subtilis* genome sequence: The molecular blueprint of a soil bacterium. *FEMS Microbiology Ecology, 28*, 1–9.

Yan, X., Yu, H.-J., Hong, Q., & Li, S. P. (2008). Cre/*lox* system and PCR-based genome engineering in *Bacillus subtilis. Applied and Environmental Microbiology, 74*, 5556–5562.

Yansura, D. G., & Henner, D. J. (1984). Use of the *Escherichia coli* Lac repressor and operator to control gene expression in *Bacillus subtilis. Proceedings of the National Academy of Sciences of the United States of America, 81*, 439–443.

Yepes, A., Schneider, J., Mielich, B., Koch, G., García-Betancur, J.-C., Ramamurthi, K. S., et al. (2012). The biofilm formation defect of a *Bacillus subtilis* flotillin-defective mutant involves the protease FtsH. *Molecular Microbiology, 86*, 457–471.

Zeigler, D. R., Pragai, Z., Rodriguez, S., Chevreux, B., Muffler, A., Albert, T., et al. (2008). The origins of 168, W23, and other *Bacillus subtilis* legacy strains. *Journal of Bacteriology, 190*, 6983–6995.

Zweers, J. C., Nicolas, P., Wiegert, T., van Dijl, J. M., & Denham, E. L. (2012). Definition of the sigma W regulon of *Bacillus subtilis* in the absence of stress. *PLoS One, 7*, e48471.

CHAPTER 5

Engineering Microbial Biosensors

Lisa Goers[*,†], Nicolas Kylilis[*,†], Marios Tomazou[*,†,‡], Ke Yan Wen[*,†], Paul Freemont[*,†,1], Karen Polizzi[*,†,1]

[*]Department of Life Sciences, Imperial College London, London, United Kingdom
[†]Centre for Synthetic Biology and Innovation, Imperial College London, London, United Kingdom
[‡]Department of Bioengineering, Imperial College London, London, United Kingdom
[1]Corresponding authors. e-mail address: p.freemont@imperial.ac.uk; k.polizzi@imperial.ac.uk

1 INTRODUCTION

In the widest sense, a biosensor is any, at least partially, biological entity that can be used to monitor a parameter of interest or a target molecule. Organisms naturally monitor their environment and react accordingly. Detection and sensing are universal in biological processes and the study of any reaction or process requires a way to monitor the process. Conversely, once we can monitor a process, we can start thinking about changing it.

Humans have always made use of such indicators, originally by drawing conclusions from the behaviour and characteristics of various organisms, such as the use of canaries in a coal mine to detect poisonous gases. With the advent of molecular biology techniques, it became possible to create nonnatural systems designed to sense and respond to the presence of metabolites and to overproduce proteins for use as recognition elements in biosensors, leading to a vast expansion in the number of biosensors currently in use.

The targeted construction and use of biosensors started in the mid-twentieth century when techniques for immobilising enzymes onto solid supports were first developed (Wingard & Ferrance, 1991). The discovery of bioluminescent (e.g. luciferase) and fluorescent (e.g. green fluorescent protein, GFP) proteins provided a convenient mechanism for visualising gene expression and protein synthesis.

Biosensors now come in many forms, including proteins attached to electrodes, promoters in genetic constructs with reporter genes or responsive fluorescent proteins. Some biosensors specifically detect certain compounds; others act as indicator species by monitoring general cell growth and wellbeing (Belkin, 2003; Darwent, Paterson, McDonald, & Tomos, 2003).

Biosensors are one of the early and most successful accomplishments in the young field of synthetic biology, which aims to harness nature's toolbox by fusing molecular biology and engineering disciplines (Boyle & Silver, 2009; van der Meer

& Belkin, 2010). Here, the discussion will concentrate on those types of biosensors that fall into the remit of synthetic biology in that they are constructed by manipulating the DNA of chassis organisms and monitoring the concentration of a particular compound or the activity of a specific responsive element.

Biosensors have a number of advantages over other detection methods, such as electrochemical signal generation, which are summarised in several of the reviews of this field (Belkin, 2003). Biosensors are *specific*. Most biosensors involve the binding of the target molecule to a protein active site. Such interactions can be extremely specific due to the carefully evolved shape of the active site through natural selection. Biosensors can also be cheap. Bacterial cells commonly used in synthetic biology often survive and grow on relatively cheap growth media and adding a biosensor circuit to a cell does not usually change the growth requirements. Such cells can easily be stored frozen and only a single cell is needed to recreate a whole population, as cells are self-repairing, self-replicating and robust and require no lengthy and expensive purification procedures as do enzymes or antibodies (Boyle & Silver, 2009; Chappell & Freemont, 2011). Through the process of gene expression, it is possible to link the detection of a target molecule by a biosensor cell to an active response by the cell.

There are also limitations to the use and applicability of biosensors, some of which can already be overcome with careful design choices and others may be addressed in the future as the field advances. Some of these are also limitations of genetic engineering in general and not just specific to biosensors. As biosensors are made from, and often located within, biological systems, there can be problems with noise, crosstalk, metabolic burden, stochastic cell behaviour, current lack of understanding and context dependence (Purnick & Weiss, 2009; Kitney & Freemont, 2012). The possible contexts for using biosensors can be limited. Cells and proteins can only operate within certain physical parameter ranges. Biosensors for certain targets may be difficult to construct. Threshold concentrations need to be considered, as biological systems tend to operate at micro- and nanomolar scales. Measuring biosensor output is only ever an indirect measure of the parameter of interest (Kelly et al., 2009). When using whole-cell biosensors, there is a risk of contamination and containment is necessary. Manipulation and engineering of biological materials usually requires high-level technical skills and technology that are currently limited to academic and industrial research labs.

2 AREAS OF APPLICATION

Biosensors are commonly used to detect the presence of particular clearly defined compounds, for example, pollutants in the environment (Wang, Barahona, & Buck, 2013) or disease indicators in the blood or physical factors such as pH (Grover et al., 2012) or presence of light (Levskaya et al., 2005). However, the potential use of biosensors in answering biological questions goes much further than this. In theory, biosensors could be constructed for any parameter and context imaginable. Parameters can include the levels of molecules such as proteins, nucleic acids and metabolites.

Through clever selection of the parameters to monitor, biosensors can be used to monitor complex biological processes and answer questions about them.

The bacterial SOS system has been successfully used to construct biosensors for several kinds of cellular stress including genotoxicity (Belkin, 2003; van der Meer & Belkin, 2010). Many projects in the iGEM (International Genetically Engineered Machine) student synthetic biology competition have the aim of producing a biosensor. Projects can be viewed on the main iGEM website (http://igem.org/Main_Page) or in project databases made by teams, for example, the Bio! iGEM History Database (http://biospot10.blogspot.co.uk/2011/09/bio-igem-history-database.html). This database contains entries for work on sensors for a huge number of different inducers, including lead, iron, light, toluene, biofilm formation, mercury, glucose, radiation, copper and cyanide. Notable iGEM biosensor projects include sensors for waterborne parasites (http://2010.igem.org/Team:Imperial_College_London) and arsenic (http://2006.igem.org/wiki/index.php/University_of_Edinburgh_2006) and pigment sensor outputs (http://2009.igem.org/Team:Cambridge).

Once constructed, a biosensor can be used as part of a genetic circuit to fill the role of an inducible element. Such regulatory elements that can be 'ON' or 'OFF' or give a concentration-dependent response are essential to build up complexity in both natural and synthetic systems (Boyle & Silver, 2009). A transcriptional or translational biosensor of some description forms an integral part of many synthetic biology genetic circuits, such as logic gates, toggle switches, counters and oscillators (Friedland et al., 2009; Purnick & Weiss, 2009).

2.1 What to target?

Many biological processes can be reduced to a small number of changing parameters, the monitoring of which can be used to study the overall process. The list of possible parameters that could be used as biosensor targets is endless and each comes with its own challenges. The detailed context of the problem gives the specifications for the development of the biosensor (Gulati et al., 2009). To illustrate, if the target is a biological molecule, one needs to consider the following properties:

a. *Location*—Is the target molecule intra- or extracellular to the biosensor cells? Are there transport mechanisms for getting the target molecule into the cells? The location of the target will define the necessary localisation of the biosensor if it is to detect the target and may restrict the type of biosensor that can be used.
b. *Concentration dynamics*—Is the target concentration likely to be fairly constant or does it change over time and, in the latter case, what is the likely dynamic range? Is the target in any way involved in the metabolism of our potential biosensor cells (van der Meer & Belkin, 2010)? This defines the range in which the sensor will need to operate to be able to detect the target.
c. *Relevance*—Is the compound unique to the process being monitored? Does the concentration of this compound really reflect the process that we want to monitor? Biosensors themselves only provide indirect measures of the parameter

of interest, so it is important to ensure that the target parameter provides relevant information (Kelly et al., 2009).
   d. *Specificity*—Is the target compound unique or quite similar to other compounds? Is there potential for interference?
   e. *Need*—Are there any current detection methods for that compound? What advantages could a biosensor bring over those existing methods?

It may be possible to find the information needed for this in the literature, or experiments may be needed to further define the parameters.

2.2 Finding potential sensors

When a suitable target has been identified, the challenge is to design a sensor that is specific to that target. As our ability to design such sensors from scratch is limited, it is usually preferable to adapt a sensor from an existing natural mechanism that is responsive to the target of interest. It is possible to take certain essential elements from a promoter and incorporate them into existing well-studied promoters (Purnick & Weiss, 2009). A literature review can be used to find naturally responsive elements. Online databases of metabolic pathways and genes, for example, (http://ecocyc.org/) and (http://www.genome.jp/kegg/), can be used to find molecules that interact with the target compound. The Registry of Standard Biological Parts (http://partsregistry.org/Main_Page) holds the parts produced in iGEM projects and a number of academic labs and includes many responsive elements. However, overall, only a small number of responsive promoters have been well characterised. Genome sequence searches could be used to identify novel responsive promoters.

Currently, only a limited number of organisms are well enough established for use as chassis in synthetic biology, but potential sensor mechanisms can be found in any organism. However, using organisms that are not commonly worked on or transferring systems between species can be challenging.

Natural sensor mechanisms tend to rely on interactions between molecules. As natural sensing mechanisms are often used to trigger gene expression in response to a stimulus, sensors fall into the categories of transcription-based, translation-based or posttranslational. Each of these categories comes with its own advantages and challenges. These will be discussed in more detail in the next section.

3 TYPES OF BIOSENSORS

Both natural and synthetic systems are regulated at each stage of the central dogma of molecular biology: transcription, RNA processing, translation and protein–protein and protein–substrate interactions (Boyle & Silver, 2009). Almost all biosensors are based on inducers acting on a stage of protein synthesis or on a completed protein. Biosensors (and other control elements) are usually distinguished by which stage of protein synthesis they act upon (Figure 5.1).

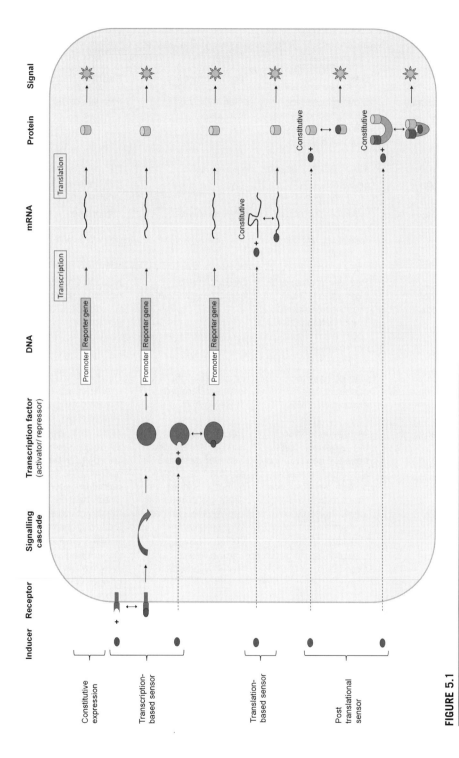

FIGURE 5.1

Mechanisms of different types of biosensors. The major types of biosensors discussed in this chapter are transcription-based, translation-based and posttranslational. In all biosensors, detection of a target molecule leads to a signal, but different biological mechanisms connect the two. (See color plate.)

Table 5.1 Characteristics of Different Classes of Biosensors

	Transcription-Based	Translation-Based	Posttranslational
Response time	Slow (minutes to hours), though there are exceptions (Purnick & Weiss, 2009)	Intermediate (minutes)	Fast (seconds)
Target location	Intra- or extracellular (receptor)	Intracellular	Intra- or extracellular (depends on protein localisation)
Form	Inducible promoters, cell-surface receptors, signalling proteins	RNA switches	Immobilised enzymes, multi-domain fusion proteins, surface-displayed proteins
Metabolic burden on the cell	Reporter protein only expressed in the presence of inducer. Transcription factor could be constitutively expressed	Need to make constitutive mRNA	Need to express constitutive protein
Ease of construction	Can be easy to construct using existing plasmids	Engineering of DNA is straightforward these days	Can be difficult to construct, as may require protein engineering
Specificity	As the recognition of the target is protein-based, it can be very specific. The final output may not be a direct measure of the target molecule due to downstream steps in protein synthesis	Target recognition can be very specific with well-designed aptamer. The final output is a more direct measure of the target molecule than for transcription-based sensors	As the recognition of the target is protein-based, it can be very specific. The final output is a more direct measure of the target molecule than for translation-based sensors
Output	Can be linked to gene expression	Can be linked to protein synthesis	Cannot be linked to gene expression. In case of FRET, fluorescence output only

These differences in mechanism lead to a diversity in the characteristics of the various biosensor classes (summarised in Table 5.1).

3.1 Response characteristics

Certain characteristics of the response are useful to define for future use of the biosensor. Different biosensors may have very different sorts of responses. The response

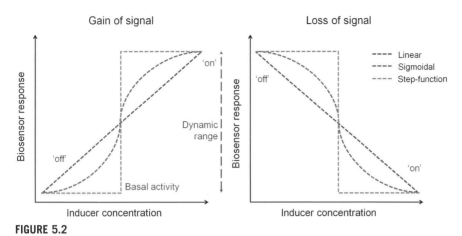

FIGURE 5.2

Idealised possible biosensor response characteristics. (For colour version of this figure, the reader is referred to the online version of this chapter.)

could come in the form of a gain of signal or a loss of signal. In cellular control mechanisms, regulators such as transcription factors can act as activators or repressors or dual-function activator–repressors. For example, the widely used AraC transcription factor of the arabinose operon acts as a repressor in the absence of its cognate inducer arabinose and as an activator in the presence of arabinose (Ogden, Haggerty, Stoner, Kolodrubetz, & Schleif, 1980). The use of different types of regulators in biosensors will lead to different output signal dynamics as shown in Figure 5.2.

The different possible response dynamics make biosensors important elements in constructing biological synthetic Boolean logic gates (Wang et al., 2013). An example of a study using sensors with positive and negative responses is by Behjousiar, Kontoravdi, and Polizzi (2012), as shown in Figure 5.3.

The dynamic range of the biosensor is described by the use of two parameters: the basal activity (the amount of activation in the absence of the molecule of interest) and the activation ratio (the highest level of activation achieved). The larger the dynamic range of the biosensor, the better its performance because it is easier to distinguish between signal and noise.

The response of the biosensor to the target molecule may, within a certain range, be a linear, sigmoidal or step function (Figure 5.2). If concentration measurements are required, then a linear response is ideal. In practice, most biosensors have a limited range over which the linear relationship will hold. Below this range, the concentration of the target is not sufficient to activate the biosensor response sufficiently to be distinguished from its basal activity. Above the range, the signal might be saturated such that further addition of the target will not result in increased output. Therefore, it is important to ensure that the linear range of the biosensor operation is matched to the expected concentration of the target in a system. The concentrations of different biological molecules that may act as targets vary hugely, as do the binding affinities of systems that may act as detectors. For biosensors in which a yes/no

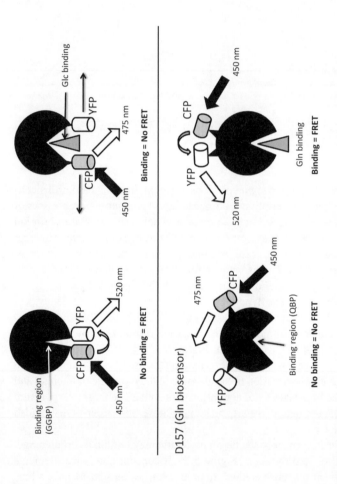

FIGURE 5.3

FRET sensors for glucose and glutamine showing positive and negative correlations with substrate concentration. (A) Schematics showing two configurations of FRET biosensors. The glucose (glc) biosensor has maximum FRET in the absence of glucose and increasing concentrations of glucose cause a decrease in signal. The glutamine biosensor (gln) shows increasing signal with increasing concentration of glutamine.

(Continued)

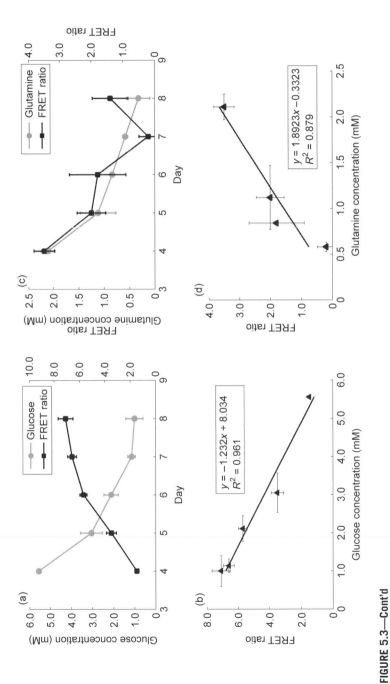

FIGURE 5.3—Cont'd

(B) The corresponding data from biosensor calibration. FRET ratios are correlated to the actual concentration of target for later use.

Figure from Behjousiar et al. (2012).

response is required, it is usually desired to increase the sensitivity of the biosensor as much as possible so that small amounts of target can be detected.

4 REPORTERS

A biosensor needs to link detection of a target to an observable output signal. The most common types of signal are fluorescence, bioluminescence and colour change. The most suitable choice will depend on the context in which the biosensor is to be used. Reporter genes and their uses for synthetic biology biosensors have been reviewed in French, Haseloff, and Ajiokaa (2011).

4.1 Fluorescence

GFP and other fluorescent proteins are widely used in molecular biology. These proteins emit fluorescence at one wavelength when exposed to another wavelength of light. They offer an output that can be detected with great specificity and sensitivity. Several fluorescent proteins can be used in the same experiment to allow monitoring of several parameters in parallel and they allow very intuitive visualisation. The use of more than one fluorescent protein is fundamental to Förster resonance energy transfer (FRET)-based biosensors. Being derived from natural proteins, they can be produced by cells and their genes can be incorporated into gene circuits as 'reporter genes'. This makes them suitable outputs for transcription- and translation-based biosensors. Many molecular biology techniques exist for measuring fluorescence.

4.2 Bioluminescence

Proteins such as firefly luciferase are similar to fluorescent proteins in their use and advantages for the engineering of biosensors, although the mechanism by which a detectable output is produced differs. Here, the protein itself does not emit light, but it acts as an enzyme that catalyses a reaction that leads to light emission. This means that the reaction substrate needs to be provided, although in some cases, the genetic construct can be modified as well to allow the cells to biosynthesise the substrate (French et al., 2011).

4.3 Colour change

For any sensor that is intended to be used 'in the field' (e.g. in a field test kit or in a hospital at point of care), an output that can be detected without the use of expensive equipment is desirable, making a colour change more suitable for such contexts than fluorescence or bioluminescence. This is a more varied category than the earlier-mentioned and includes a number of very different systems. Most of such systems require the addition of a substrate that is chemically transformed into a related colour-producing compound. A commonly used reporter that causes a colour change is the *lacZ* (β-galactosidase) gene and the chromogenic compound 5-bromo-4-chloro-3-indolyl-β-D-galactopyranoside (Xgal). The 2010 Imperial College London iGEM team used the compound catechol and the enzyme catechol 2,3-dioxygenase (C23O) to produce a yellow colour (http://2010.igem.org/Team:Imperial_College_London). The 2009 Cambridge iGEM team made a set of 'colour generators' using pigments (http://2009.igem.org/Team:Cambridge). The 2006 Edinburgh iGEM team made an arsenic sensor with an output that resulted in a pH change, which could be visualised using a coloured pH indicator (http://2006.igem.org/wiki/index.php/University_of_Edinburgh_2006).

5 BIOSENSORS AND SYNTHETIC BIOLOGY

The biosensors discussed here are constructed through systematic design and engineering of genetic circuits using a synthetic biology approach. Synthetic biology aims not only to engineer biological parts, devices and systems but also to facilitate the process of engineering biology and to make it more predictable (Kitney & Freemont, 2012). The approach includes the concepts of modularity and standardisation. In a biosensor context, this means that we want to be able to construct our genetic circuits as far as possible from preexisting standard parts that can be assembled fast and efficiently. We also want to have the option to exchange individual parts to optimise or adapt the sensor for specific contexts. The following sections will explain in more detail how to construct different classes of biosensors using these principles.

6 TRANSCRIPTION-BASED BIOSENSORS

Transcription-based biosensors make use of promoters controlled by responsive transcription factors or signalling pathways. This section will describe step-by-step how to design, construct and test such biosensors and the considerations necessary at each stage.

6.1 Planning the construct

Having identified a suitable responsive promoter, it needs to be incorporated into a gene circuit. While gene circuits of different biosensors will have context-dependent differences, there are some common elements (Figure 5.4). The construct needs to contain a selective marker (usually an antibiotic resistance gene) and the potential biosensor promoter needs to be put in a context where its behaviour can be observed, that is, where it can be active. Required for this is a reporter gene flanked by a ribosome binding site (RBS) and a transcription terminator. Examples of biosensor genetic constructs are also shown by van der Meer and Belkin (2010).

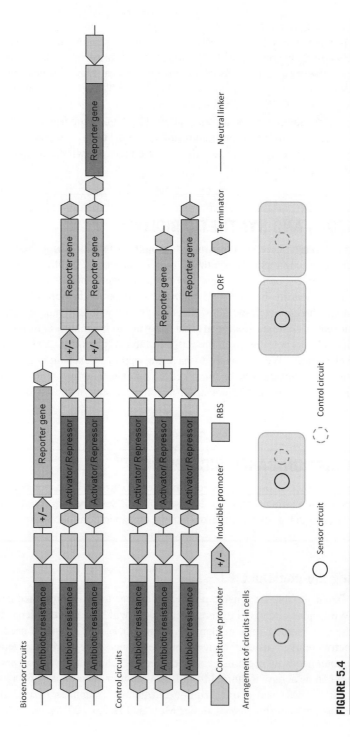

FIGURE 5.4

Typical elements of biosensor and control cell gene circuits and possible arrangements of biosensor and control circuits in the cells. ORF stands for open reading frame and refers to any gene coding sequence. (See color plate.)

Overexpression of the relevant transcription factor will pose an additional metabolic burden but is often necessary. If the transcription factor comes from another organism, it needs to be expressed as part of the sensor construct. If the transcription factor is endogenous to the sensor cells, it may not need to be overexpressed. However, if the reporter plasmid is present at a high copy number, then a single endogenous copy of the transcription factor may not give optimal sensor output levels.

Construction of a negative control construct or cell line can aid the proper characterisation of the biosensor. Not all studies use genetic circuits to construct a negative control cell line. For example, Kelly et al. (2009) used an untransformed wild-type cell line as the negative control measurement when measuring promoter activity through GFP expression using a flow cytometer. However, they also introduced the idea of a promoter reference standard for promoter characterisation, hence using ratiometric relative measurements instead of absolute measurements to account for differences in cellular conditions at different experimental conditions.

Ratiometric analyses are ubiquitous in biology and medical sciences, for example, the body-mass index, which measures body weight while taking into account height. This principle can be applied to biosensor design and testing by expressing a second reporter gene under the control of a reference promoter. FRET sensors are inherently ratiometric in their use. Ratiometric measurements have advantages and disadvantages (Kelly et al., 2009). They can take account of various kinds of experimental variation. However, ratiometric measurements only give relative and not absolute measurements and absolute quantities are often necessary for computational modelling. It may be possible to use conversion factors to convert relative measurements to absolute measurements, but this may reintroduce the experimental variation that ratiometry was used to avoid in the first place.

The parts needed to construct the genetic circuit can be obtained from a number of sources. There are many commercially available vectors that contain the parts needed to express constructs of interest. Specific parts of interest can be requested from other academic labs or synthesised directly. The Registry of Standard Biological Parts (http://partsregistry.org/Main_Page) holds the parts produced in iGEM projects and a number of academic labs. All the parts are constructed using the BioBrick standard. However, there are limitations to this registry and there are plans in place for a more professional part registry (Kitney & Freemont, 2012).

There is software available to assist with the construction of certain parts of certain aspects of designing genetic circuits. For designing RBSs and libraries thereof, there exists an online calculator (https://www.denovodna.com/software/).

Many computational tools for design and analysis of genetic circuits have been listed and reviewed elsewhere (Purnick & Weiss, 2009; Kitney & Freemont, 2012). *In silico* testing of constructs can help select from several possible circuit designs (Gulati et al., 2009).

6.2 Assembling the construct

Once the genetic circuit design exists on paper, it is time to construct the DNA vector and transform it into an appropriate chassis. While there are many staple DNA assembly techniques in molecular biology, assembly remains a likely limiting step in biosensor engineering and synthetic biology in general. Common assembly methods include restriction enzyme-based methods, PCR-based methods or gene synthesis by commercial companies (Ellis, Adie, & Baldwin, 2011; Kitney & Freemont, 2012). DNA assembly methods using microfluidic platforms have also been developed (Szita, Polizzi, Jaccard, & Baganz, 2010).

The behaviour of bioparts is context-dependent, meaning that parts arranged in a different order within a gene circuit may behave differently. Different constructs of different orders can be constructed through combinatorial assembly.

Libraries of promoters, RBSs or other parts can be created to search sequence space for different characteristics, for example, tune strength. This can be done by commercial companies or in-house using error-prone PCR or other directed evolution methods.

6.3 Testing the construct

After completion of biosensor construction, the next step is to characterise the sensor experimentally.

6.3.1 What to measure?

In 2010, van der Meer and Belkin reviewed the general types of experiments that could be done to characterise the different aspects of biosensors (see Figure 5.5). These include measuring biosensor output over time, target concentration or distance from target or presence of other compounds. These experiments thus give information on aspects such as response time, dose–response, range and specificity of the biosensor.

6.3.2 How to measure?

The type of data that need to be collected to characterise any particular biosensor is largely determined by the choice of reporter. Commonly used techniques for visualising biosensor output signals include spectrophotometers/plate readers (fluorescence, bioluminescence and absorbance), flow cytometers (fluorescence) and microscopy (fluorescence and bioluminescence).

All of fluorescence, bioluminescence and colour change can be monitored using standard plate reader equipment that usually incorporates both fluorescence and ultraviolet/visible light spectrophotometer capabilities. Measuring optical density at 600 nm gives an indication of cell density, which can be used to calculate the biosensor output per cell or synthesis rate of the reporter protein. Using flow cytometry or microscopy, the signal output of single cells can also be monitored closely. Examples of studies using flow cytometry include that of Kelly and colleagues (2009).

To avoid some of the lengthy assembly and transformation steps and reduce biological noise for initial characterisation or selection experiments, cell-free *in vitro* methods can be used (Chappell, Jensen, & Freemont, 2013).

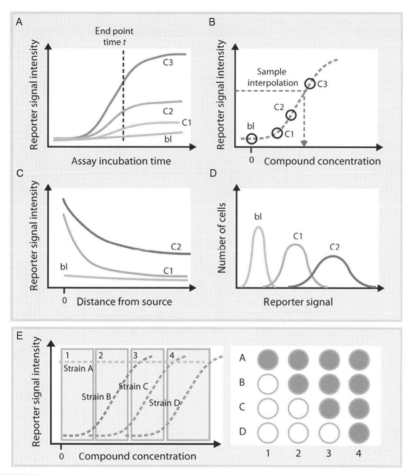

FIGURE 5.5

Types of measurement that may be performed in bioreporter assays. (A) Measurement of reporter output over time for several different target concentrations compared with a noninduced control (C1, C2 and C3 are the three different target concentrations and bl (baseline) as the control). (B) Assay endpoint measurements at time t as a function of different target concentrations; target concentrations in unknown samples are quantified by interpolating their signal in the reporter assay using the calibration curve for pure compound. (C) Reporter output as a function of distance to a target source. (D) Single-cell measurements; in this case, signals from all of the individual cells are plotted as distribution curves from which an average response can be calculated. (E) 'Traffic light bioreporter systems' that can work independently of external calibration assays. This assay uses, for example, four isogenic strains with the same reporter output but differing in the compound concentration threshold at which their reporter circuit is activated. The number of reporter strains reacting to a sample (rather than their reporter signal intensity *per se*) is then representative for the compound concentration range. (See color plate.)

Reprinted with permission from Macmillan Publishers Ltd., van der Meer and Belkin (2010). Copyright (2010).

Most of the previously mentioned techniques lend themselves to integration with high-throughput platforms, such as robotics or microfluidics (Gulati et al., 2009; Szita et al., 2010). Robotics allows automation of processes using machinery and computers. Microfluidics is the manipulation of tiny amounts of fluids and it allows reduced sample volume and parallel analysis of large number of samples (Szita et al., 2010). As the technique already builds on the detection of fluorescent reporter proteins, it is very suited to this context. Microfluidic platforms can be used for cell-free assays (Szita et al., 2010). High-throughput techniques and techniques for use 'in the field' have been previously reviewed (Belkin, 2003; van der Meer & Belkin, 2010).

There have been attempts in synthetic biology towards establishing standards of biopart characterisation, but this has lagged behind the establishment of assembly standards (Canton, Labno, & Endy, 2008; Kelly et al., 2009). Absolute values of experimental results can vary across experimental conditions and measurement instruments, as all types of cellular biosensors are strongly influenced by overall cell behaviour (Kelly et al., 2009). One way to get around this that has been explored is the use of a reference standard (Kelly et al., 2009).

OpenWetWare (http://openwetware.org) is an online resource for the synthetic biology community where details of many experimental protocols can be found. For example, the protocol used in the multi-institution promoter activity experiment by Kelly et al. (2009) can be found in (http://openwetware.org/wiki/The_BioBricks_Foundation:Standards/Technical/Measurement/Promoter_characterization_experiment_(FACS)).

Overall, the design of characterisation experiments and methods of data analysis will depend on construct design. During data analysis, an assessment will need to be made as to whether the promoter in question acts as a suitable biosensor. As can be seen in Figure 5.2, ideally, the biosensor should have a low-output baseline, that is, in the absence of inducer, there should be no output. Promoters that do allow gene expression in the absence of inducer are said to have 'leaky expression'. There needs to be a suitably large difference between the 'ON' and 'OFF' states of the sensor, which will depend on the context of the biosensor used.

7 TRANSLATION-BASED BIOSENSORS

As an alternative to transcription-based biosensing systems, RNA switches have been engineered to respond to the presence of a target of interest. In this case, the RNA molecules are always present in the cell and the response results from a change in RNA processing or translation in the presence of the target. In general, they have three functional domains for receiving the input signal, signal processing and producing a measurable output (Figure 5.6):

- In RNA switches, the signal receiver (input) is an aptamer domain. An aptamer domain is an RNA motif that is able to recognise and bind a target of interest specifically. This recognition is the result of the specific secondary and tertiary structure formed by the RNA molecule in the presence of the target.

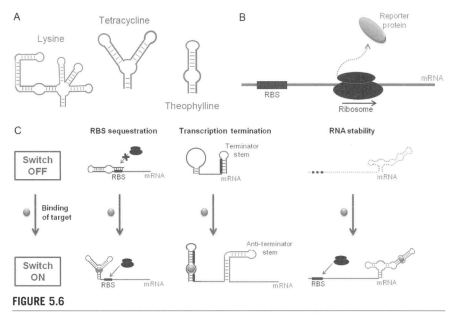

FIGURE 5.6

The three functional modules that constitute an RNA switch. (A) The signal receiver (aptamer). (B) The output signal domain generates reporter protein expression through translation. (C) The three mechanisms of signal processing at the molecular level: RBS sequestration, premature termination and modification of RNA stability. (For colour version of this figure, the reader is referred to the online version of this chapter.)

- Signal processing is achieved through the switch domain that responds to target binding by modulating the processing, translation or stability of the device.
- RNA switches return the output from signal processing by the expression of a reporter protein, the level of which corresponds to the number of switches activated in a positive or negative manner.

RNA-based systems can offer a number of advantages over the transcription-based systems discussed previously. These include faster response times, decreased metabolic burden on cells, the fact that RNA folding is universal and there is an abundance of computer-aided design (CAD) tools for RNA (Liu & Arkin, 2010).

7.1 Planning the construct

Aptamers are the signal input domain of the construct. They are DNA or RNA molecules that can specifically bind a target molecule. Target recognition stems from their ability to form complex three-dimensional shapes. Aptamers are predominantly unstructured in solution, but folding upon association with the target provides the opportunity for hydrogen bonding, stacking interactions, shape complementarity and electrostatic complementarity (Hermann & Patel, 2000).

RNA aptamer sequences have been isolated *in vitro* for a very large range of targets, from small molecules such as nucleotides, amino acids, dyes, cofactors and chemicals to larger molecular entities such as proteins, protein complexes and whole cells. Researchers can access the sequence of existing aptamers online from original research papers or aptamer databases such as Aptamer Base (http://aptamerbase.semanticscience.org/). Alternatively, if an aptamer for a particular target of interest does not already exist, it can be developed by the SELEX (Systematic Evolution of Ligands by Exponential Enrichment) method (Ellington & Szostak, 1990) or ordered from a private company that offers this service.

A riboswitch is an mRNA molecule that is able to self-regulate its translation in response to the presence of a small molecule. Most natural riboswitches occur in bacteria and are employed to sense a diverse class of metabolites such as inorganic ligands (metals and anions), purines, cofactors of proteins and amino acids. They often function as feedback control systems for the production of proteins and enzymes involved in the biogenesis or transport of the metabolites they sense (recently reviewed in Serganov and Nudler (2013)).

A number of natural riboswitches with a defined input and output signal are available and can be exploited for the purposes of constructing a biosensor. However, at times, it will be necessary to construct a switch for a target where a natural switch does not exist. Researchers inspired by nature have come up with a number of strategies for engineering new RNA switches. One common strategy is the sequestration/release of an RBS by a change in secondary structure of the mRNA molecule in presence or absence of the target (Figure 5.6). For example, Lynch, Desai, Sajja, and Gallivan (2007) engineered an RNA switch responsive to theophylline using a library generation and high-throughput screening approach. Alternatively, the Smolke lab engineered a riboswitch that functions via transcript stabilisation/destabilisation (Win & Smolke, 2007). These are only two examples of a large number of strategies that can be devised. Other strategies have been used by Qi, Lucks, Liu, Mutalik, and Arkinn (2012) and Ceres, Garst, Marcano-Velázquez, and Bateyy (2013).

The output domain of an RNA switch is constituted by the RBS and the coding sequence of the desired reporter protein. The repertoire of reporter proteins has been discussed previously.

7.2 Riboswitch construction guidelines

This section is intended to serve as a guide for the engineering of a bacterial *cis*-acting RNA switch device as described by Lynch and Gallivan (2009). The device can accommodate a user-defined input and output signal and its switch function is achieved by the mechanism of RBS sequestration. The resulting device will exhibit a low signal in the absence of the target and a higher signal when the molecule is present. This construction method utilises a library of DNA sequences between the aptamer and the RBS, followed by high-throughput screening (fluorescence-activated cell sorting (FACS) or other methods) to identify functional devices.

This type of switch mechanism can be explained by a thermodynamic model. The RNA transcript can exist in two different secondary structure conformations: the 'OFF' and 'ON' states. In the absence of the target, the 'OFF' state is thermodynamically favoured due to lower free energy. In the presence of the target, its binding to the aptamer domain releases energy and shifts the equilibrium of the secondary structure of the transcript to the 'ON' state. Thus, the difference between the free energy of the secondary structures of the 'OFF' and 'ON' states must be less than the free energy of the binding of the target to the aptamer domain.

The starting point is the assembly of the DNA sequence encoding the RNA switch. The construct is very similar to a transcription-based biosensor with one addition, which is the insertion of an RNA aptamer between the promoter and the RBS that will serve to control the translation (Figure 5.7). It is useful to use a fluorescent reporter so that FACS can be used in the selection of functional switches. As before, any DNA assembly method can be selected for the purposes of construction.

For library creation, the method of inverse PCR is employed to randomise the area between the aptamer and the RBS to yield a series of sequence variants. The original DNA plasmid is used as the template and synthesised primers with overhangs of randomised nucleotides are employed to generate diversity. The number of nucleotides randomised can be anywhere from 4 to 12, with a library of 8 randomised nucleotides producing a library size of 65,536 members.

To aid the screening process, FACS can be employed to select functional devices by using two sets of sorting (Figure 5.8; Lynch & Gallivan, 2009). First, the library of mutants is transformed into cells and is screened using FACS in the absence of the target. The portion of this library that exhibits the lowest fluorescence is selected and retained. This identifies switches in which the sequence between the aptamer and the RBS prevents translation. This population is then cultured to expand the number of cells and subjected to FACS again. This time, the population is assessed both in the presence and absence of the target. The cells that exhibit high fluorescence in the presence of the target (higher than the population in the absence of the target, which serves as a baseline) are retained. By now, a small, manageable number of variants should have been isolated. These can be assessed individually for significant differences in translation initiation rate in absence/presence of the target to identify switches with the best performance.

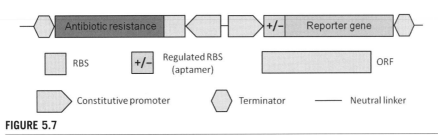

FIGURE 5.7

Typical elements of an RNA switch device. (For colour version of this figure, the reader is referred to the online version of this chapter.)

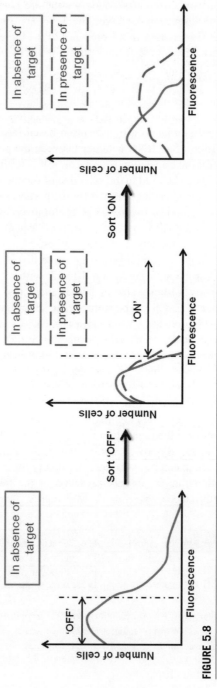

FIGURE 5.8

FACS-assisted screening for a functional, high-quality RNA switch device. Successive sorting experiments are used to isolate switches that are nonfluorescent in the absence of the target molecule and fluorescent in its presence using the procedure outlined in Lynch and Gallivan (2009). (For colour version of this figure, the reader is referred to the online version of this chapter.)

8 POSTTRANSLATIONAL BIOSENSORS

Posttranslational biosensors are based on proteins that are constitutively produced but become activated in the presence of the target, leading to the generation of a signal. One classical example of a posttranslational biosensor is an enzyme that catalyses the reaction of the target with one or more secondary substrates to produce a signal when the target is present. The signal generated is then proportional to the concentration of the target. This model has been used in constructing glucose biosensors for diabetes monitoring (Yoo & Lee, 2010). Because of the diversity of protein behaviours available, the design of posttranslational biosensors is much less generic than with transcription- or translation-based biosensors.

One interesting class of posttranslational biosensors for intracellular monitoring in the context of synthetic biology is protein switches based on Förster resonance energy transfer (FRET; reviewed in Frommer, Davidson, and Campbelll (2009)). FRET is a nonradiative transfer of energy between two fluorophores, where the emission from one fluorophore (donor) excites the second fluorophore (acceptor) causing an increase in emission from the acceptor (Figure 5.3). The fluorophores in question can be chemical or biological (e.g. fluorescent proteins). In the latter case, the biosensor can be genetically encoded and produced constitutively within the cells, readily available to measure the target. This type of biosensor can be used to measure the intracellular concentration of a molecule of interest quantitatively (Okumoto, Jones, & Frommer, 2012) and therefore is of interest for synthetic biology and metabolic modelling.

8.1 Planning the construct

Constructing a FRET-based posttranslational biosensor requires constructing a fusion protein, which is composed of two fluorophores and at least one protein domain that binds the target (Figure 5.3A). In addition to the requirement for two fluorophores with overlapping spectra, FRET also relies on the proximity and orientation of the fluorophores. Therefore, the fusion must be arranged in such that when the protein folds into its three-dimensional structure, there is a difference in the proximity and/or orientation of the fluorophores when the target is present versus when it is absent. This will, in turn, lead to a change in the amount of energy transfer (the detectable signal), which is dependent on the concentration of the target.

Identifying appropriate protein domains that bind the target is the hardest task in constructing a FRET-based posttranslational biosensor. This is because not only it is sufficient for the domain to bind the molecule but also the binding event must result in a conformational change that can be exploited to detect the binding event. Protein structures in the presence and absence of the target (e.g. from the Protein Data Bank; http://www.rcsb.org/pdb/home/home.do) can be useful to the guide design. Certain classes of proteins are also known for their large conformational changes upon binding. For example, bacterial periplasmic binding proteins undergo a hinging motion when binding their intended targets. This has been exploited for a number of designed sensors (Dwyer & Hellinga, 2004).

Finally, for concentration measurements, it is necessary for the linear range of detection of the sensor to correspond to the expected concentrations of the target within the cell. This can be modulated by making mutants of the binding domain with increased or decreased affinity for the target, depending on the concentrations expected.

8.2 Assembling the construct

The arrangement of the fusion protein will greatly impact the sensor function. A linear DNA sequence encoding the donor fluorophore, the target binding domain and the acceptor fluorophore in sequence is unlikely to provide a sensor with good characteristics. Instead, it may be necessary to insert the gene for one of the fluorophores into the coding region of the target binding domain and attach the other fluorophore to one of the termini (e.g. Bogner & Ludewig, 2007) or to make truncations from the N- or C-termini of some of the coding regions (e.g. Deuschle et al., 2005) in order to better exploit the conformational change of the resulting protein. It may also be necessary to use a second interaction domain that recognises the molecule binding domain with the target bound and in turn undergoes a conformational change that magnifies the response (e.g., Miyawaki et al., 1997). Bioinformatics analysis of sequences can aid in this process (Pham, Chiang, Li, Shum, & Truong, 2007). However, many FRET-based posttranslational biosensors are currently constructed using library generation techniques followed by high-throughput screening to identify sensors with the best signal-to-noise ratio.

It will also be necessary to include the genetic elements necessary to ensure the biosensor is produced within the cells such as a promoter, RBS and selection marker as with the other types of biosensors (Figure 5.9). For convenience, a constitutive promoter can be used to produce the biosensor so that no additional steps are necessary to make the measurements. The strength of the RBS should be sufficient to produce a strong fluorescent signal, but not so strong as to cause a high metabolic burden on the cells from protein production.

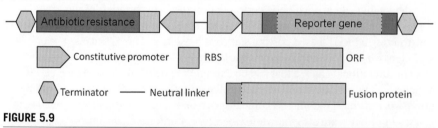

FIGURE 5.9

Typical elements of a posttranslational biosensor. The protein is produced constitutively within the cell and reports changes in target concentration by emitting a signal. (For colour version of this figure, the reader is referred to the online version of this chapter.)

8.3 Testing the construct

The measure of output of a FRET-based posttranslational biosensor is the FRET ratio. Data can be collected using fluorescence microscopy and confocal microscopy or in a 96-well plate format using whole cells or purified protein and with the addition of different concentrations of the target. The samples are excited at the maximum excitation wavelength of the donor fluorophore and the emission is monitored at two wavelengths: the emission maximum of the donor (corresponding to energy that was not transferred to the second fluorophore) and that of the acceptor (energy that was transferred, resulted in excitation of the second fluorophore, and was emitted). The two sets of emission data (arbitrary units) are then divided to calculate the FRET ratio (Figure 5.10). Figure 5.3 shows two examples for metabolite biosensing. Thus, FRET ratios are an inherently ratiometric measurement and differences in expression level of the biosensor are (within a reasonable range) normalised away.

9 *IN VITRO* BIOSENSORS

Although the biosensors discussed so far are designed for a cellular chassis, it is also possible to apply the same principles to develop an *in vitro* biosensor. Cell-free systems have long been considered a useful tool for studying biological processes or for synthesising proteins, and this ability to maintain fundamental transcription/translation reactions outside of an encircling membrane can also be harnessed for the detection of target molecules. *In vitro* systems are either synthetic enzymatic pathways (SEPs), in which the required system is reconstituted from its individual parts, or crude extract cell-free systems (CECFs), which use the products of cell lysis with minimal further processing (Hodgman & Jewett, 2012). The latter system is far more common as it is extremely difficult and expensive to construct SEPs.

There are a number of advantages that *in vitro* cell-free systems offer in comparison to whole-cell chassis. The primary characteristic of cell-free systems is a reduction in complexity. Since they are pared down to a few fundamental biological processes, the available resources in each individual reaction are focused only on the parts that have been purposely included by the designer (Simpson, 2006). Living cells on the other hand have their own metabolic and regulatory processes that can conflict with the functioning of a synthetic biosensor. Another benefit is that some biosensors may contain parts that can only exist in a cell-free context. For example, detecting certain targets may require expression of a protein that would lead to loss of viability for their host cells (Harris and Jewett, 2012) or the biosensor could be targeting a molecule that is unable to diffuse across cell membranes. Finally, since cell-free systems are not classified as genetically modified organisms, they may face less strict safety regulation and therefore be more suited for the transition to applications outside of controlled laboratory settings (Forster & Church, 2007). However, cell-free systems are less suitable for large-scale applications when taking into consideration the cost of preparing cell-free extracts versus the relatively cheap process of

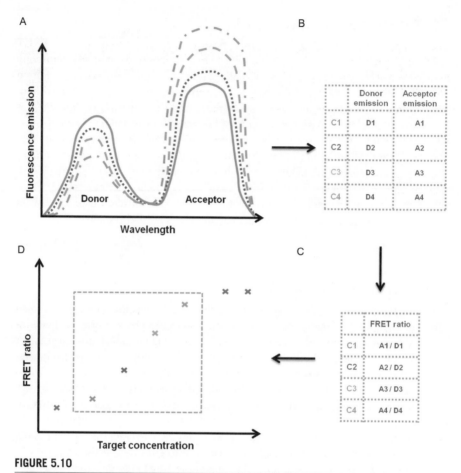

FIGURE 5.10

Measurement of FRET ratios. (A) The fluorescence emission spectrum (arbitrary units) is measured. (B) The emission at the maxima of the donor and acceptor fluorophores is recorded. (C) The emission of the acceptor fluorophore is divided by that of the donor fluorophore to obtain a FRET ratio. (D) A plot of the FRET ratio versus concentration can be used to interpolate the concentration of target in a sample. The box depicts the linear range of measurement where the output of the biosensor is most reliable. (See color plate.)

cellular self-replication. Furthermore, the implementation of complex biological circuits in *in vitro* settings is still in its early stages and has yet to be widely demonstrated (Karig, Iyer, Simpson, & Doktycz, 2012; Shin & Noireaux, 2012).

A minimal *in vitro* biosensor can be constructed with a few basic components. These are (i) cell-free extracts produced from crude cell lysate, (ii) a reaction buffer containing amino acids and energy sources and (iii) a DNA template encoding the gene circuit required for biosensor functionality. The latter can encode any type of biosensor discussed in the previous sections. CECFs can be prepared experimentally or obtained commercially—*E. coli,* wheat germ or rabbit reticulocyte lysate are among the most common sources (Hodgman & Jewett, 2012). The DNA template can be purified from the cells, but this requires lengthy *in vivo* procedures such as transformation and cell culture. In order to reduce preparation time, DNA can also be generated entirely *in vitro* using a modified PCR and ligation protocol (Chappell et al., 2013). Once constructed, the biosensor output can be detected with the same methods as described previously, for example, fluorescence spectrophotometers.

10 MODELLING BIOSENSORS

Developing a corresponding mathematical model can provide potential users with the necessary information about the characteristics and performance of a biosensor. Furthermore, the model can inform the design and experimental implementation of the physical biosensor, since a model-guided rational design of the choice of RBSs, promoters or gene variants and network connectivity can be a much more efficient process than any 'trial-and-error' approach.

10.1 What is a model and how to plan the modelling process

In a wider context, a model is anything that can represent the physical system theoretically. Models in biology usually appear as sets of chemical or mathematical equations. A gene or a protein normally interacts with multiple other molecules within the cell that can lead to a large degree of complexity if all species are modelled in detail. Therefore, decisions must be made as to which information is vital to include. The result is a model that is an abstract approximation of the system of interest but hopefully retains sufficient information to be informative about the behaviour of the system. Even simplified biochemical models are often multivariable and non-linear, making them impossible to solve analytically. Instead, encoding them in a computational framework allows the use of numerical solvers to predict within certain confidence levels the trajectory and functionality of a specific system.

The choice of model type and its resolution depends on the amount of information available for model development and the minimum detail needed to render all the critical functions of the system. The model can be either a mechanistic model that explicitly simulates all the biochemical reactions underlying its operation or something as simple as a transfer function that relates the concentration of the target molecule to the biosensor output. The former can describe both the function and the structure of the physical network and even predict complex behaviour of the system that can be verified experimentally. It can be used to explore possible modifications and their implications on the biosensor characteristics. For example, fine-tuning a biosensor to be more sensitive or more robust can be supported by a mathematical model that allows testing promoters of different transcriptional strength.

Other forms of simpler models will render a similar function but without explicitly demonstrating the structure. These types of models, also known as phenomenological models, can be created empirically when prior knowledge of the system behaviour exists (i.e. derived from a dataset) and fitted with more detailed and diverse experimental data. Alternatively, they can be derived from a mechanistic model by introducing assumptions that effectively reduce and dedimensionalise the model. The objective here is to reduce the number of variables and parameters that must be estimated. This allows for further mathematical manipulation, not applicable in a large-scale model, able to highlight some aspects or a critical parameter of the system.

Based on their mathematical properties, models can be categorised into different groups, such as deterministic, stochastic, continuous or discrete and linear or non-linear. Additional details on the function of different types of models are beyond the scope of this chapter, but the interested reader can find more information in several published sources (Sontag, 2005; Bayer et al., 2012; Stan, 2013).

In either case, the model should capture the basic input–output relationships of the physical biosensor both qualitatively and quantitatively. In simple terms, a model should answer as accurately as possible the following question:

- If the biosensor for target L is exposed to X concentration of L, how strong will the response signal be?

$$Y = f(X) \tag{5.1}$$

Alternatively, the question should also be answerable the other way around:

- For a response that amounts to X units of reporter output, what is the concentration of target L in the sample?

$$X = f(Y) \tag{5.2}$$

A more detailed model can hold much more information like the following:

- How fast is the response after exposing the biosensor to the target?
- For how long I can safely measure the output after the initial exposure?
- What kind of response will I get when I alternate the spatiotemporal characteristics of my input signal?
- Is the system robust or will it exhibit a radically different input–output relationship with a slight change of environmental conditions or parameters?
- What will change if I replace a specific part with a different variant?
- How sensitive and specific is the sensor?

$$Y(t) = f(X, t, \ldots) \tag{5.3}$$

Hence, for each individual project, the questions that need to be addressed must be identified prior to the modelling process so that an appropriate level of detail is incorporated into the theoretical representation. The goal is to avoid developing unnecessarily large-scale models that are hard to manipulate and compute but at the same time avoid over-simplified models that neglect critical aspects of the operation of the biosensor.

A general guideline for the modeller can be found in Stan (2013). Briefly, it is useful to identify the following:

1. Which parts of the system are important? This can be, but is not limited to, the input/output substances and the sensory molecules.
2. What can be measured or monitored during the operation of the system, for example, the amount of GFP protein or an intermediate fluorescent metabolite?
3. What can be used to optimise the sensor, for example, an RBS, a promoter or a degradation tag?
4. The range of conditions of operation and possible limitations that the environment will possibly impose on the system.

The earlier-mentioned provides a first look into what needs to be represented in the final model.

10.2 Constructing the model

The modelling procedure normally goes through multiple iterations until a model that is able to capture the necessary dynamics in a reliable way is reached. Typically, the models are derived directly from the main biochemical reactions that are taking place in the specific system.

For example, assume that a receptor, R, binds to its target, L, at a rate of k_1 and forms a complex (RL). The receptor and target can either dissociate with a rate of k_{-1} or bind and activate a promoter (G) that controls the expression a fluorescent protein GFP. This is a simple type of biosensor (depicted in Figure 5.11). The equations that describe this system are as follows:

$$R + L \underset{k_{-1}}{\overset{k_1}{\rightleftarrows}} RL \tag{5.4}$$

$$RL + G_0 \underset{k_{-2}}{\overset{k_2}{\rightleftarrows}} G_1 \tag{5.5}$$

$$G_1 \overset{k_t}{\longrightarrow} GFP + G_1 \tag{5.6}$$

Furthermore, the target and the receptor are introduced to the system by a rate of k_l and k_p, respectively, and degrade/diffuse at a rate of k_d:

$$R \underset{k_p}{\overset{k_d}{\rightleftarrows}} \varnothing \tag{5.7}$$

$$L \underset{k_l}{\overset{k_d}{\rightleftarrows}} \varnothing \tag{5.8}$$

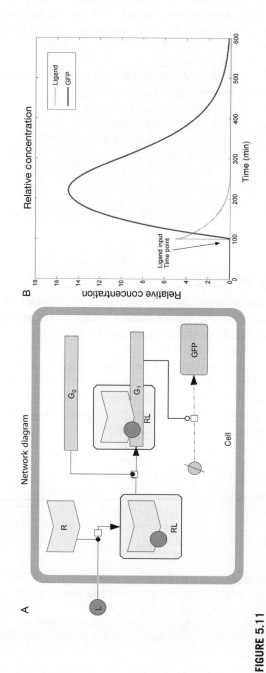

FIGURE 5.11

A simple biosensor circuit built with CellDesigner™ (Table 5.3) and a time-course simulation of the dynamics when L is injected into the system at $t = 100$ min.

$$\text{GFP} \xrightarrow{k_\text{d}} \varnothing \qquad (5.9)$$

$$\text{RL} \xrightarrow{k_\text{d}} \varnothing \qquad (5.10)$$

$$\text{G}_1 \xrightarrow{k_\text{d}} \text{G}_0 \qquad (5.11)$$

Breaking this simple model down into the component parts, we can identify a set of species (types of molecules) and rates. The rates and other constants are used to make up the set of parameters. All of the species are floating in a confined space, which can be either a cell or a subcellular vesicle. In the example so far, there are

- species: R, L, RL, GFP, G_0, G_1,
- parameters: k_1, k_{-1}, k_2, k_{-2}, k_t, k_d, k_p, k_l,
- containers: cells.

By applying the law of mass action, a set of ordinary differential equations (ODEs) that represent the corresponding mathematical model can be derived:

$$\frac{d[\text{R}]}{dt} = k_{-1}[\text{RL}] + k_\text{p} - k_1[\text{R}][\text{L}] - k_\text{d}[\text{R}] \qquad (5.12)$$

$$\frac{d[\text{L}]}{dt} = k_{-1}[\text{RL}] + k_\text{l} - k_1[\text{R}][\text{L}] - k_\text{d}[\text{L}] \qquad (5.13)$$

$$\frac{d[\text{RL}]}{dt} = -k_{-1}[\text{RL}] + k_1[\text{R}][\text{L}] - k_\text{d}[\text{RL}] - k_2[\text{G}_0][\text{RL}] + k_{-2}[\text{G}_1] \qquad (5.14)$$

$$\frac{d[\text{GFP}]}{dt} = k_\text{t}[\text{G}_1] - k_\text{d}[\text{GFP}] \qquad (5.15)$$

$$\frac{d[\text{G}_0]}{dt} = -k_2[\text{G}_0][\text{RL}] + k_{-2}[\text{G}_1] + k_\text{d}[\text{G}_1] \qquad (5.16)$$

$$\frac{d[\text{G}_1]}{dt} = k_2[\text{G}_0][\text{RL}] - k_{-2}[\text{G}_1] - k_\text{d}[\text{G}_1] \qquad (5.17)$$

10.3 Model reduction

Model reduction is an essential process in order to obtain a model with fewer variables, less dimensions and fewer nonlinearities. From the example in the preceding text, it is apparent that even for a very simple system, the corresponding mechanistic model can become very complicated. It is worth pointing out that even this is still an abstract representation, the availability of resources (amino and nucleic acids) and processes like transcription, translation and protein folding are all 'embedded' in the reactions and not explicitly taken into account.

Nevertheless, the mathematical model can be reduced by introducing assumptions, provided that these are supported by data from the literature. The two most commonly used reduction approaches in biological models are as follows:

1. Introduction of conservation laws: for example, assuming that the number of genes G_t either free or bound to RL is constant,

$$\frac{d[G_t]}{dt} = \frac{d[G_0]}{dt} + \frac{d[G_1]}{dt} = 0 \Rightarrow G_t = G_0 + G_1 \qquad (5.18)$$

2. Quasistationary (or quasisteady state) approximations (QSA): Given certain conditions, it is valid to assume that some reactions happen much faster than others, such that some species reach their equilibrium almost instantly compared to other species. In this case, the concentration of these species does not change. This difference of reaction speed is known as time-scale separation. In the example here, it can be assumed that the formation of the G_1 complex happens on the scale of microseconds while the other species react in milliseconds; therefore, the rate of change of G_1 is essentially zero:

$$\frac{d[G_1]}{dt} = k_2[G_0][RL] - k_{-2}[G_1] - k_d[G_1] = 0,$$
$$\Rightarrow [G_1] = \frac{k_2[G_0][RL]}{k_{-2} + k_d} \qquad (5.19)$$

Substituting Equation (5.19) into Equations (5.14) and (5.15) and (5.16), G_1 can be then be eliminated.

Introducing the same assumption for RL, substituting and manipulating algebraically, Equation (5.12) can be rewritten as

$$\frac{d[GFP]}{dt} = \frac{a}{b}[R][L] - k_d[GFP] \qquad (5.20)$$

where a and b are a combination of the parameters. Equation 5.20 is a direct correlation of input to output (L to GFP) and the available free receptor, which is ideal for predicting the behaviour, but not easily achievable for more complex models.

3. Other methods include dedimensionalisation by parameter elimination and QSA for quantities of different scales (e.g. an excess amount of species L will effectively result into a rate of change close to zero).

For more information about model reduction, the reader can refer to the literature (De Jong, 2002; Sontag, 2005; Bayer et al., 2012; Stan, 2013).

10.4 Identifying the starting conditions and parameter values

This is a critical and challenging task in order to reach a truly representative model. Before any kind of simulation is possible, the model needs to be fed with actual

Table 5.2 Databases for Models, Parameters and Bioinformatic Tools

BioNumbers	http://bionumbers.hms.harvard.edu/
BioModels Database	http://www.ebi.ac.uk/biomodels-main/
CellML Model Repository	http://models.cellml.org/cellml
The Cell Collective	http://www.thecellcollective.org
KEGG	http://www.genome.jp/kegg/
BioCyc	http://biocyc.org/
ExPASy	http://www.expasy.org/

values regarding the starting concentration of each species and parameter. Ideally, each parameter should be experimentally determined; however, this is often impossible due to time and resources. In reality, experimental measurements are usually limited to the species and parameters that are most important and easily measured.

For sources of the nonexperimentally determined values, the modeller can refer to the following:

- Literature. Any published material about the specific genes or constructs that are used in the biosensor is a potential source of parameters or ratios of parameters such as k_d (dissociation constants) of a target from a protein.
- Repositories and databases of models or parameters for different organisms (Table 5.2).
- Statistical approximations. If a parameter value cannot be found in the literature, a value that falls within a biologically acceptable range can be a valid starting point.
- Parameter fitting. If an experimental dataset is available, parameters can be fit so that the resulting trajectory of the model matches the data as closely as possible. This fitting can only be performed for a limited number of parameters and only when most of the other values have been determined.
- In models with a large number of undetermined parameters it is possible to realise multiple parameter sets that satisfy the data. However, caution should be taken with parameters values that are far from realistic values.

More information on parameter inference can be found in Ashyraliyev, Fomekong-Nanfack, Kaandorp, and Blomm (2009), MacDonald, Barnes, Kitney, Freemont, and Stan (2011) and Bayer et al. (2012).

10.5 Making use of the model—numerical simulations

Although the example model is relatively small in size and simple in structure, it is still a nonlinear model. Nonlinear models cannot generally be solved analytically, but they can be integrated (solved) numerically. Most of the modern ODE solvers are based on Euler's method, where instead of solving for the continuous space of the independent variable (time in our case), it is possible to make a fairly good

Table 5.3 GUI Model Editors and Tools	
SimBiology® 3 (MATLAB® package)	http://www.mathworks.co.uk/products/simbiology/
Systems Biology Workbench	http://sbw.sourceforge.net/
TinkerCell	http://www.tinkercell.com/
CellDesigner™	http://www.celldesigner.org/
COPASI	http://www.copasi.org/

approximation over a short discrete point of time (or a time step, dt) (Butcher, 1987). The smaller the step, the more accurate the prediction will be, but the trade-off is that smaller steps require more calculations. Computational models can be written in most programming languages, but it is more convenient to use those that have built-in numerical solvers such as MATLAB® or R. Alternatively, there are a variety of graphical user interface (GUI) tools and model editors that can be used to write, manipulate and simulate models (Table 5.3). SimBiology® 3.0 in MATLAB® is an excellent example, since it supports many automated functions, works with most of the existing solvers in MATLAB® (including stochastic ones) but still allows the user direct access to the code so that it can be customised for new applications. The Systems Biology Markup Language (SBML) (Hucka et al., 2003) is another open-source project that delivers software packages that automate most of the basic processes, provide solvers and are also connected to online databases and repositories of models, rate constants and genetic parts databases.

The tasks performed can include time-course simulations, scanning the response difference over a range of values for a specific variable (Figure 5.12A and B) or even simulation of an event (e.g. a sudden increase of input signal). Finally, a sensitivity analysis (Figure 5.12C) can reveal critical parameters to focus on and investigate further. Sensitivity analysis is based on quantifying the amount of influence an individual parameter value has over the output. Those with a high degree of influence are important to developing a good model and may require direct measurement in order to create a reliable model.

10.6 Validating the model and reiterating the modelling procedure

A model cannot be perfect; hence, all of the earlier-mentioned steps can be revisited in order to create a model that matches the experimental system more closely. The validity of assumptions may be reviewed, leading to a less or a more reduced version. In addition, new components that may affect the function of a system can be introduced. Finally, a model can be converted into its stochastic counterpart in order to account for the variability. The best models should be able to make valid predictions for the behaviour of a sensor at conditions different than the ones that were used to fit experimental parameters. Other validation criteria that can be used are the accuracy, predictability, reusability and parsimony (level of abstraction) of the final model (MacDonald et al., 2011).

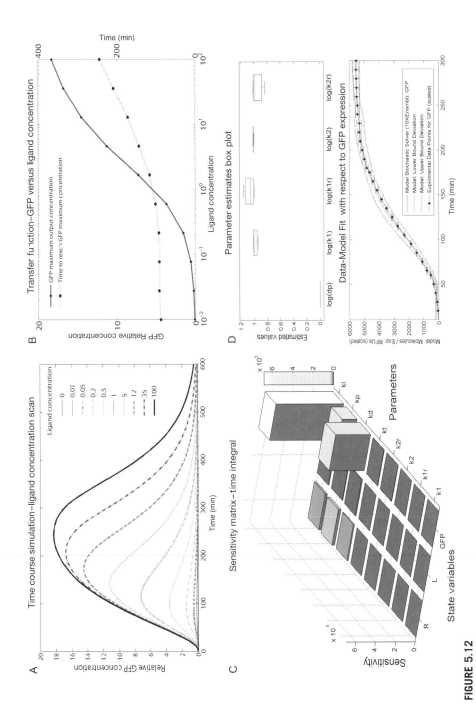

FIGURE 5.12

Simulations illustrating how the model can be used: (A) a time-course simulation for different concentrations of target (L), (B) the predicted transfer function relating target concentration and the maximum GFP output and the time it takes to reach that point. (C) Sensitivity analysis reveals that the rate of change of GFP is affected more strongly by the rate of target addition (k_t) and the degradation rate (k_d). (D) Fitting model parameters with experimental data.

11 OUTLOOK

By applying a synthetic biology workflow, it is now possible to streamline the development of new biosensors and to make more sophisticated constructs that can be used to detect complex phenomena. However, we are still limited by existing biological knowledge. For example, to develop new biosensors for the diagnosis of complex diseases such as cancer, we still lack indicators of the disease, which are both specific and universally reliable. The problem often lies with finding a suitable target to detect that is indicative of the process to be observed. This problem might potentially be solved by the use of logic gates to link multiple inputs to create an integrated response. These principles can also be used to build more complex systems using biosensors that allow communication or decision-making. Steps towards this have already been taken (Wang et al., 2013). Since logic gates can be designed in a modular format, the principles of synthetic biology will still hold.

However, while simpler logic gates have been constructed, more complex systems that integrate a larger amount of information have not yet been made. In order to achieve a logic gate that assimilates many inputs, an increased understanding of natural gene regulation is required. Even though we currently have a reasonably good understanding of certain individual mechanisms (e.g. specific operons), we do not yet know how the expression of many different genes is integrated overall. In order to build complex biosensors and synthetic biology constructs composed of many components, an understanding of the system's complexity would be useful. On the other hand, making such synthetic constructs might also provide insights into how the natural systems are regulated.

To make biosensors applicable to a wider range of target concentrations and to enhance biosensor response, new mechanisms useful for signal enhancement could be developed. For example, signalling cascades, rather than single transcription events, could be employed to magnify the signal and to help overcome biological noise by improving the signal-to-noise ratio (van der Meer & Belkin, 2010). In addition, 'rapid response' circuits can be developed in order to decrease the response time of transcription-based biosensors (http://2010.igem.org/Team:Imperial_College_London). For both translation-based and posttranslational biosensors, the development of sensors for new targets is still a bottleneck as there is a lack of generic, modular technologies for construction. What is needed in this case is a series of platform technologies into which sensor specificity could be designed rapidly as the need for new biosensors is identified. For RNA switches, steps towards this have been made. It was recently shown that more modular RNA switches could be developed from eukaryotic internal ribosome entry sites by adding additional RNA domains that have complementarity to each other and remove the context dependence of the aptamer domain by insulating it from the rest of the switch (Ogawa, 2012).

Another challenge is to begin to deploy the biosensors outside of the research laboratory. Despite work in whole-cell biosensors that dates back several decades, the number of biosensors that have been used in the field is very limited (van der Meer & Belkin, 2010; French et al., 2011). This might be partially due to issues surrounding the containment of genetically modified organisms. If so, the expansion of cell-free systems and systems to prevent unwanted gene transfer between cells could help overcome this issue. In addition, small portable devices based on microfluidics could offer opportunities for contained use in the field (e.g. Prindle et al., 2012).

Given the potential benefits of harnessing biology for detection purposes, applying synthetic biology methods to the development of biosensors should enable the development of cost-effective detection methods for a number of important applications. This includes diagnostic tools and mechanisms to help researchers decipher complex biological processes. Ideally, to move the field forward, in the future, we will need to be able to increase the complexity of targets to make them more representative of the processes of interest and be able to rapidly develop new sensors from scratch without appropriating existing systems that might be subject to crosstalk within the systems.

References

Ashyraliyev, M., Fomekong-Nanfack, Y., Kaandorp, J. A., & Blom, J. G. (2009). Systems biology: Parameter estimation for biochemical models. *FEBS Journal*, *276*(4), 886–902.

Bayer, T. S., Baldwin, G., Dickinson, R. J., Ellis, T., Freemont, P. S., Kitney, R. I., et al. (2012). *Synthetic biology: A primer*. London: Imperial College Press, Singapore: World Scientific.

Behjousiar, A., Kontoravdi, C., & Polizzi, K. M. (2012). In situ monitoring of intracellular glucose and glutamine in CHO cell culture. *PLoS One*, *7*(4), e34512.

Belkin, S. (2003). Microbial whole-cell sensing systems of environmental pollutants. *Current Opinion in Microbiology*, *6*(3), 206–212.

Bogner, M., & Ludewig, U. (2007). Visualization of arginine influx into plant cells using a specific FRET-sensor. *Journal of Fluorescence*, *17*(4), 350–360.

Boyle, P. M., & Silver, P. A. (2009). Harnessing nature's toolbox: Regulatory elements for synthetic biology. *Journal of the Royal Society Interface*, *6*(Suppl. 4), S535–S546.

Butcher, J. C. (1987). *The numerical analysis of ordinary differential equations: Runge–Kutta and general linear methods*. Chichester: Wiley-Interscience.

Canton, B., Labno, A., & Endy, D. (2008). Refinement and standardization of synthetic biological parts and devices. *Nature Biotechnology*, *26*(7), 787–793.

Ceres, P., Garst, A. D., Marcano-Velázquez, J. G., & Batey, R. T. (2013). Modularity of select riboswitch expression platform enables facile engineering of novel genetic regulatory devices. *ACS Synthetic Biology*, *2*(8), 463–472. http://dx.doi.org/10.1021/sb4000096.

Chappell, J., & Freemont, P. S. (2011). Synthetic biology—A new generation of biofilm biosensors. In *Paper presented at the The Science and Applications of Synthetic and Systems Biology: Workshop Summary*.

Chappell, J., Jensen, K., & Freemont, P. S. (2013). Validation of an entirely in vitro approach for rapid prototyping of DNA regulatory elements for synthetic biology. *Nucleic Acids Research*, *41*(5), 3471–3481.

Darwent, M. J., Paterson, E., McDonald, A. J. S., & Tomos, A. D. (2003). Biosensor reporting of root exudation from Hordeum vulgare in relation to shoot nitrate concentration. *Journal of Experimental Botany*, *54*(381), 325–334.

De Jong, H. (2002). Modeling and simulation of genetic regulatory systems: A literature review. *Journal of Computational Biology, 9*(1), 67–103.

Deuschle, K., Okumoto, S., Fehr, M., Looger, L. L., Kozhukh, L., & Frommer, W. B. (2005). Construction and optimization of a family of genetically encoded metabolite sensors by semirational protein engineering. *Protein Science, 14*(9), 2304–2314.

Dwyer, M. A., & Hellinga, H. W. (2004). Periplasmic binding proteins: A versatile superfamily for protein engineering. *Current Opinion in Structural Biology, 14*(4), 495–504.

Ellington, A. D., & Szostak, J. W. (1990). In vitro selection of RNA molecules that bind specific ligands. *Nature, 346*, 818–822.

Ellis, T., Adie, T., & Baldwin, G. S. (2011). DNA assembly for synthetic biology: From parts to pathways and beyond. *Integrative Biology, 3*(2), 109–118.

Forster, A. C., & Church, G. M. (2007). Synthetic biology projects in vitro. *Genome Research, 17*(1), 1–6.

French, C. E., Haseloff, J., & Ajioka, J. (2011). Synthetic biology and the art of biosensor design. In *Paper presented at the The Science and Applications of Synthetic and Systems Biology: Workshop Summary*.

Friedland, A. E., Lu, T. K., Wang, X., Shi, D., Church, G., & Collins, J. J. (2009). Synthetic gene networks that count. *Science Signaling, 324*(5931), 1199.

Frommer, W. B., Davidson, M. W., & Campbell, R. E. (2009). Genetically encoded biosensors based on engineered fluorescent proteins. *Chemical Society Reviews, 38*(10), 2833–2841.

Grover, A., Schmidt, B. F., Salter, R. D., Watkins, S. C., Waggoner, A. S., & Bruchez, M. P. (2012). Genetically encoded pH sensor for tracking surface proteins through endocytosis. *Angewandte Chemie International Edition, 51*(20), 4838–4842.

Gulati, S., Rouilly, V., Niu, X., Chappell, J., Kitney, R. I., Edel, J. B., et al. (2009). Opportunities for microfluidic technologies in synthetic biology. *Journal of the Royal Society Interface, 6*(Suppl. 4), S493–S506.

Harris, D. C., & Jewett, M. C. (2012). Cell-free biology: exploiting the interface between synthetic biology and synthetic chemistry. *Current Opinion in Biotechnology, 23*, 672–678.

Hermann, T., & Patel, D. J. (2000). Adaptive recognition by nucleic acid aptamers. *Science, 287*(5454), 820–825.

Hodgman, C. E., & Jewett, M. C. (2012). Cell-free synthetic biology: Thinking outside the cell. *Metabolic Engineering, 14*(3), 261–269.

Hucka, M., Finney, A., Sauro, H. M., Bolouri, H., Doyle, J. C., Kitano, H., et al. (2003). The systems biology markup language (SBML): A medium for representation and exchange of biochemical network models. *Bioinformatics, 19*(4), 524–531.

Karig, D. K., Iyer, S., Simpson, M. L., & Doktycz, M. J. (2012). Expression optimization and synthetic gene networks in cell-free systems. *Nucleic Acids Research, 40*(8), 3763–3774.

Kelly, J. R., Rubin, A. J., Davis, J. H., Ajo-Franklin, C. M., Cumbers, J., Czar, M. J., et al. (2009). Measuring the activity of BioBrick promoters using an in vivo reference standard. *Journal of Biological Engineering, 3*(1), 4.

Kitney, R., & Freemont, P. (2012). Synthetic biology—The state of play. *FEBS Letters, 586*(15), 2029–2036.

Levskaya, A., Chevalier, A. A., Tabor, J. J., Simpson, Z. B., Lavery, L. A., Levy, M., et al. (2005). Synthetic biology: Engineering Escherichia coli to see light. *Nature, 438*(7067), 441–442.

Liu, C. C., & Arkin, A. P. (2010). Cell biology. The case for RNA. *Science, 330*(6008), 1185–1186.

Lynch, S. A., Desai, S. K., Sajja, H. K., & Gallivan, J. P. (2007). A high-throughput screen for synthetic riboswitches reveals mechanistic insights into their function. *Chemistry & Biology, 14*(2), 173–184.

Lynch, S. A., & Gallivan, J. P. (2009). A flow cytometry-based screen for synthetic riboswitches. *Nucleic Acids Research, 37*(1), 184–192.

MacDonald, J. T., Barnes, C., Kitney, R. I., Freemont, P. S., & Stan, G.-B. V. (2011). Computational design approaches and tools for synthetic biology. *Integrative Biology, 3*(2), 97–108.

Miyawaki, A., Llopis, J., Heim, R., McCaffery, J. M., Adams, J. A., Ikura, M., et al. (1997). Fluorescent indicators for Ca^{2+} based on green fluorescent proteins and calmodulin. *Nature, 388*(6645), 882–887.

Ogawa, A. (2012). Rational construction of eukaryotic OFF-riboswitches that downregulate internal ribosome entry site-mediated translation in response to their ligands. *Bioorganic & Medicinal Chemistry Letters, 22*(4), 1639–1642.

Ogden, S., Haggerty, D., Stoner, C. M., Kolodrubetz, D., & Schleif, R. (1980). The Escherichia coli L-arabinose operon: Binding sites of the regulatory proteins and a mechanism of positive and negative regulation. *Proceedings of the National Academy of Sciences of the United States of America, 77*(6), 3346–3350.

Okumoto, S., Jones, A., & Frommer, W. B. (2012). Quantitative imaging with fluorescent biosensors. *Annual Review of Plant Biology, 63*, 663–706.

Pham, E., Chiang, J., Li, I., Shum, W., & Truong, K. (2007). A computational tool for designing FRET protein biosensors by rigid-body sampling of their conformational space. *Structure, 15*(5), 515–523.

Prindle, A., Samayoa, P., Razinkov, I., Danino, T., Tsimring, L. S., & Hasty, J. (2012). A sensing array of radically coupled genetic 'biopixels'. *Nature, 481*(7379), 39–44.

Purnick, P. E., & Weiss, R. (2009). The second wave of synthetic biology: From modules to systems. *Nature Reviews Molecular Cell Biology, 10*(6), 410–422.

Qi, L., Lucks, J. B., Liu, C. C., Mutalik, V. K., & Arkin, A. P. (2012). Engineering naturally occurring trans-acting non-coding RNAs to sense molecular signals. *Nucleic Acids Research, 40*(12), 5775–5786.

Serganov, A., & Nudler, E. (2013). A decade of riboswitches. *Cell, 152*(1–2), 17–24.

Shin, J., & Noireaux, V. (2012). An E. coli cell-free expression toolbox: Application to synthetic gene circuits and artificial cells. *ACS Synthetic Biology, 1*(1), 29–41.

Simpson, M. L. (2006). Cell-free synthetic biology: A bottom-up approach to discovery by design. *Molecular Systems Biology, 2*, 69.

Sontag, E. D. (2005). Molecular systems biology and control. *European Journal of Control, 11*(4), 1–40.

Stan, G.-B. (2013). Modelling in biology—Lecture notes. http://www.bg.ic.ac.uk/research/g.stan/2010_Course_MiB_article.pdf.

Szita, N., Polizzi, K., Jaccard, N., & Baganz, F. (2010). Microfluidic approaches for systems and synthetic biology. *Current Opinion in Biotechnology, 21*(4), 517–523.

van der Meer, J. R., & Belkin, S. (2010). Where microbiology meets microengineering: Design and applications of reporter bacteria. *Nature Reviews Microbiology, 8*(7), 511–522.

Wang, B., Barahona, M., & Buck, M. (2013). A modular cell-based biosensor using engineered genetic logic circuits to detect and integrate multiple environmental signals. *Biosensors and Bioelectronics, 40*, 368–376.

Win, M. N., & Smolke, C. D. (2007). A modular and extensible RNA-based gene-regulatory platform for engineering cellular function. *Proceedings of the National Academy of Sciences of the United States of America*, *104*(36), 14283–14288.

Wingard, L. B., Jr., & Ferrance, J. P. (1991). Concepts, biological components, and scope of biosensors. In *Biosensors with Fiberoptics* (pp. 1–27). Presume New York: Springer.

Yoo, E. H., & Lee, S. Y. (2010). Glucose biosensors: An overview of use in clinical practice. *Sensors (Basel)*, *10*(5), 4558–4576.

Noise and Stochasticity in Gene Expression: A Pathogenic Fate Determinant

Mikkel Girke Jørgensen, Renske van Raaphorst, Jan-Willem Veening[1]

Molecular Genetics Group, Groningen Biomolecular Sciences and Biotechnology Institute, Centre for Synthetic Biology, University of Groningen, Groningen, The Netherlands
[1]*Corresponding author. e-mail address: j.w.veening@rug.nl*

1 INTRODUCTION

Individual cells in a bacterial population grown in the same environment never exhibit exactly the same phenotype, despite a common genetic identity. This phenomenon is known as phenotypic variation. At the genetic level, this means that individual isogenic cells in a bacterial population show variable gene expression patterns, which translate into changes in protein levels and thus influence the cellular behaviour to a certain extent. Stochastic gene expression, or gene expression 'noise', has been proposed as a major source of this variability (reviewed in de Lorenzo & Perez-Martin (1996); Locke & Elowitz (2009); Munsky, Neuert, & van Oudenaarden (2012)). This phenomenon has, until recently, received little scientific attention, in part because classical molecular microbiologists have assumed that isogenic cells respond in a nonfluctuating fashion to a given stimulus and, thus, the traditional techniques used in laboratories to study gene expression rely on pooling of millions of cells and therefore determine the average values for the entire population (Dubnau & Losick, 2006). However, with the emergence of single-cell analytical techniques such as flow cytometry and fluorescence microscopy, it has become clear that phenotypic variation is ubiquitous in nature and occurs in many biological processes.

In general, gene expression is intuitively a 'noisy' process with multiple molecular origins. While beneficial for certain traits such as virulence (see the succeeding text), noise in gene expression is altogether an unwanted by-product for the cell and all organisms, both prokaryotes and eukaryotes, tend to reduce and control it in relation to essential processes. An example of such a control mechanism is negative feedback loops (Fraser, Hirsh, Giaever, Kumm, & Eisen, 2004; Dublanche, Michalodimitrakis, Kummerer, Foglierini, & Serrano, 2006; Smits, Kuipers, & Veening, 2006). Strikingly, it was found that autoregulated systems such as self-repression show a decrease in noise levels compared to an unregulated system (Becskei & Serrano, 2000).

2 ORIGINS OF NOISE

To explain the concept of noise, we can start by considering a bacterial cell about to divide. Let us assume that a cell in this state will have all its molecules in Brownian motion. This means that all particles or molecules move randomly around the cell. Therefore, the chance that each new daughter cell will inherit the same number of ribosomes, transcription factors, RNA polymerases and so on is negligible (Munsky et al., 2012). If we scale that up, the chance that two cells out of an entire isogenic population, grown under the same conditions, are identical with respect to the number and composition of their molecules is minute. As a natural consequence of this randomness in the distribution of molecules, there will be cell-to-cell variations in most biological processes in the population. This type of variation is termed extrinsic noise.

The mechanisms causing extrinsic noise in gene expression include, for example, the concentrations of RNA polymerases and ribosomes or regulatory factors that would lead to variations in the level of expression of a given gene between one cell and another but not between two identical genes in the same cell (Swain, Elowitz, & Siggia, 2002; Raser & O'Shea, 2005). Thus, extrinsic noise arises from sources that are global to a single cell but vary from one cell to another. Intrinsic noise on the other hand arises from random fluctuations in the biochemical process of gene expression itself regardless of the presence of extrinsic noise. Let us consider a purely hypothetical population of isogenic bacterial cells, containing exactly the same number and composition of molecules. The amount of protein produced by any given gene would still fluctuate from one cell to the other (Elowitz, Levine, Siggia, & Swain, 2002). This intrinsic noise arises from randomness in the binding of transcription factors to the promoter region, of RNA polymerases to the promoter and of ribosomes to the ribosomal binding site on the messenger RNA. The rate of translation can also contribute to intrinsic noise, since the availability of tRNA for each ribosome and mRNA turnover by ribonucleases are processes that show probabilistic behaviours. The counterpart to protein synthesis, protein degradation, also adds to the pool of intrinsic noise.

Thereby, we can in principle define intrinsic noise as that arising directly from the process of gene expression and extrinsic noise as that arising from changes in the intercellular environment (Elowitz et al., 2002; Swain et al., 2002; Dublanche et al., 2006). 'Noise' can therefore be defined as variation in gene expression in a population of isogenic cells. Phenotypic variation and total noise is the sum of all these parts. A typical image of gene expression noise is seen in Figure 6.1A. Here, isogenic cells of *Streptococcus pneumoniae* are expressing *gfp*. Even a constitutive promoter generates cell-to-cell variability, which can be observed and quantified by fluorescence microscopy.

2.1 Transcriptional bursting

Recently, a set of exciting experiments with single-molecule detection has shown that gene expression occurs as transcriptional bursts. In a hallmark paper, Golding and coworkers detected individual mRNA transcripts in individual living cells of

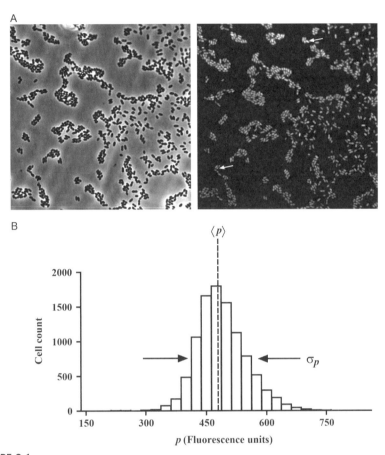

FIGURE 6.1

Image analysis and data output for measuring noise in gene expression. (A) Microscopic image acquisition of an isogenic culture of *Streptococcus pneumoniae* grown under identical conditions. The cells express GFP from a constitutive promoter; see phase contrast image (left) and GFP signal (right image). Even when grown under identical conditions, the fluorescence (and likely the protein levels) of GFP varies from one cell to another (compare the two cells marked by white arrows). (B) A typical FACS output of the cells in (A). Here, 10,000 *S. pneumoniae* cells are analysed within a few seconds. From the graphical output, it is clear that the level of GFP can be described in statistical terms such as the mean value $\langle p \rangle$ and a standard deviation σ_p.

Escherichia coli in real time (see the succeeding text; Golding, Paulsson, Zawilski, & Cox, 2005). Strikingly, the authors found that *E. coli* cells produced transcripts in short distinct bursts and not at a steady rate as would be expected according to a normal distribution (Golding et al., 2005; Raj & van Oudenaarden, 2008). Based on their data, the authors suggested that transcriptional bursting is a result of an ON/OFF model where gene expression randomly switches back and forth between active and inactive transcription. Consequently, in light of these authors' findings, a simple Poisson stochastic model of gene expression is unlikely to apply. The Poisson expression model, in which transcripts are constitutively expressed at a constant rate and degraded in a first-order reaction, is the simplest model (Li & Xie, 2011; Munsky et al., 2012). Moreover, the transcriptional bursting model observed in *E. coli* reflects findings in eukaryotic cells (Raj & van Oudenaarden, 2008). At the mechanistic level, transcriptional bursting can be explained by promoter kinetic and promoter transition states. Binding of a transcriptional activator to the promoter region, for instance, would likely lead to several RNA polymerases associating with the promoter in successive order, thereby resulting in several rounds of transcription before the activator disassociates from the promoter again. This ON/OFF switching would then result in high and low transcription rates, respectively (Kaern, Elston, Blake, & Collins, 2005). Zenklusen and coworkers found both constitutively expressed and transcriptional bursting genes using *Saccharomyces cerevisiae* as a model organism. These findings reflect a complex nature of gene expression and demonstrate several modes of transcription modulation (Zenklusen, Larson, & Singer, 2008). It should be noted that transcriptional bursting may not be a common source of noise in prokaryotes. In fact, most models concerning prokaryotic gene expression assume that the transition between ON and OFF switching is so fast that the promoters are in steady states (Kaern et al., 2005). In addition, studies in *E. coli* and *Bacillus subtilis* revealed that the stochasticity in protein production was a result of translational bursting and not transcriptional bursting (Ozbudak, Thattai, Kurtser, Grossman, & van Oudenaarden, 2002; Yu, Xiao, Ren, Lao, & Xie, 2006). Transcriptional bursting is thought to be more important in eukaryotic gene expression and the transcriptional bursting observed in bacteria is much weaker and measured only on an inducible gene (Golding et al., 2005; Zenklusen et al., 2008). However, recent single-molecule fluorescence *in situ* hybridization experiments revealed that small RNAs are often bimodally expressed in clonal bacterial populations and, in the ON cells, typically contain between 1 and 10 copy numbers, in line with a bursting model of gene transcription (Shepherd et al., 2013). In eukaryotic systems, chromatin remodelling between open and closed structures correspond to the ON and OFF promoter transition states, and it has become the prominent model for gene expression control (Blake et al., 2006; Chubb, Trcek, Shenoy, & Singer, 2006; Kaufmann & van Oudenaarden, 2007; Zenklusen et al., 2008).

3 MEASURING NOISE

In recent years, it has become possible to detect and analyse gene expression at the single-cell and single-molecule level. These advances in the experimental protocols have been pivotal in detecting and quantifying the variability of gene expression. In particular, the development of green fluorescent protein (GFP) as a reporter to count the number of molecules—either mRNA or protein—has been essential. Today, the use of GFP (or its variants), coupled to single-cell imaging, is the method of choice for studying noise in gene expression (Figure 6.1A).

Most techniques used in molecular biology require pooling of a vast number of cells, resulting in an averaging that is not representative for individual single cells (for instance, samples taken from bacterial cultures for downstream analyses such as microarray and proteomics). By studying population heterogeneity, it has become clear that each cell in a population, even when grown at identical conditions, is unique. Depicting gene expression as a single average value based on a population sample is therefore somewhat misleading and must be interpreted with care. Rather, gene expression in a population is more accurately represented by a distribution with associated statistical properties such as standard deviation and variance (Nevozhay, Adams, Van Itallie, Bennett, & Balazsi, 2012; Figure 6.1B). Fluorescence-based techniques are the most direct tools for determining gene expression noise, either by measuring protein or mRNA levels in single cells (Figure 6.1A; Larson, Singer, & Zenklusen, 2009). Today, several types of fluorophores are available to visualize cell population heterogeneity. The principal methods for the detection of fluorescent signals are *f*luorescence-*a*ctivated *c*ell *s*orting (FACS) and microscopy-based cell imagining. Both techniques measure the fluorescence emitted at the single-cell level. FACS has the advantage of analysing several thousands of cells at once (Figure 6.1B). However, cell sorting is less sensitive and does not facilitate the detection of low-abundance signals (Larson et al., 2009). Microscopy-based cell imaging, on the other hand, is more sensitive with a high dynamic range, and single cells can be tracked over time and generations (de Jong, Beilharz, Kuipers, & Veening, 2011). Data acquisition, conversely, requires several time series of pictures and fewer cells are analysed compared to FACS (Larson et al., 2009).

Measuring noise for a given gene usually involves promoter fusion to a fluorescent protein like GFP and then determining the total level of fluorescence signal per cell either by flow cytometry or microscopy. FACS analysis immediately generates a fluorescent value for each cell including statistical properties such as mean signal and standard deviation. Microscopy images on the other hand require additional analysis. Several software tools exist to quantify the level of fluorescent signal emitted by each cell from microscopic images, for example, ImageJ (Schneider, Rasband, & Eliceiri, 2012), MicrobeTracker (Sliusarenko, Heinritz, Emonet, & Jacobs-Wagner, 2011), and Schnitzcells (Young et al., 2012). Although the use of flow cytometry and microscopy in combination with a fluorescent reporter protein facilitates the monitoring of gene expression at the single-cell level, it should be borne in mind that the correct folding and maturation of these proteins themselves can also contribute to the total noise.

The detection of single proteins is a challenging task as it requires the recording of each single protein in a cell. This is made difficult as proteins diffuse within the cytoplasm making the signal spread and image acquisition problematical. In a milestone paper, Yu and coworkers overcame this difficulty by making a fusion protein of a yellow (YFP) variant of GFP to the membrane-bound Tsr protein as a reporter to monitor the *lac* promoter activity (Yu et al., 2006). The membrane anchoring slowed the diffusion rate, thereby making it possible to study single proteins in live cells by fluorescence microscopy. The use of photobleaching allowed the authors to monitor single-protein production, and they observed that proteins were produced as bursts from a single mRNA molecule (Yu et al., 2006; Larson et al., 2009).

Another way to detect noise in gene expression is by two-photon fluorescence fluctuation microscopy. In brief, this technique measures the intensity fluctuations of signals from a fluorescent protein inside a cell at each pixel in a set of fast scanned images, which are then deconvolved, allowing for counting of the molecular brightness and determination of the absolute number of fluorescent proteins diffusing inside cells. Using this novel approach, Ferguson and coworkers recently showed that a *B. subtilis* glycolytic promoter driving GFP showed strong transcriptional bursting and, surprisingly, that for highly 'bursty' promoters, negative feedback does not suppress the noise (Ferguson et al., 2012).

*F*luorescence *in situ h*ybridization (FISH) is a cytogenetic technique used to detect and quantify single mRNA (or DNA) species within a cell. FISH uses DNA probes complementary to the specific target mRNA. As the probes are conjugated to a fluorescent dye, fluorescent microscopic imaging allows the quantification and localization of the mRNA. Usually, several probes complementary to the same mRNA are used to enhance the signal-to-noise ratio (Femino, Fay, Fogarty, & Singer, 1998; Larson et al., 2009; Trcek et al., 2012). However, the cells have to be fixed, limiting the method to visualizing only the mRNA at a snapshot in time. Conversely, the MS2 system based on the RNA-binding phage protein MS2 fused to a fluorescent protein like GFP allows the trafficking of mRNA molecules in live cells. Using a reporter mRNA with multiple stem–loop structures recognized by the MS2 fusion protein provides a molecular beacon, which can be visualized by fluorescent microscopy. The method benefits from using living cells as the mRNA reporter molecule is tracked and quantified (Golding et al., 2005; Querido & Chartrand, 2008; Trcek et al., 2012).

Recently, a set of RNA aptamers have been designed that bind fluorophores (Paige, Wu, & Jaffrey, 2011). These RNA–fluorophore complexes can be used to visualize and localize mRNAs in live cells. One such RNA–fluorophore complex, called Spinach, closely resembles the fluorescence properties of GFP (Paige et al., 2011) and such RNA-adapter approaches are rapidly becoming a popular combination to study RNA dynamics in live cells (Armitage, 2011) and we foresee that they will be used to study noise in gene expression.

4 ENGINEERING NOISE

Noise in gene expression has long been predicted and several studies have quantified, measured and modelled noise in many genetic systems. The field of research has expanded rapidly and attracted scientists ranging from geneticists and biophysicists to theoretical mathematicians. From a system biologist's point of view, measuring and quantifying noise is not so much the goal as to seek to model and engineer the level of noise. Constructing robust gene circuits is a challenging task and optimizing a gene network usually requires minimizing heterogeneity (Kaern, Blake, & Collins, 2003). Also, for biotechnological applications, it might be desirable to reduce heterogeneity in production of a commercially valuable molecule, for instance.

Several parameters are useful when discussing phenotypic variability and gene expression noise. The distribution of gene expression of a single gene can be described by a mean value of expression denoted $\langle p \rangle$ with a standard deviation σ_p; see Figure 6.1B. The relative standard deviation $\sigma_p/\langle p \rangle$ is sometimes used as a measure of noise. However, the Fano factor ($\sigma_p^2/\langle p \rangle$), or phenotypic noise strength, is a more commonly used measurement of noise. This is because the relative standard deviation changes as the mean value changes, whereas the phenotypic noise strength is less sensitive to changes in the mean value. The Fano factor is thus a noise measurement that directly correlates with the width of the population distribution (Thattai & van Oudenaarden, 2001; Ozbudak et al., 2002; Kaern et al., 2003). Another important measure is the coefficient of variance, $\sigma_p^2/\langle p \rangle^2$, and this is important in relation to engineering of noise since it gives a measure of the signal-to-noise ratio (Kaern et al., 2003).

In a landmark paper, Ozbudak and colleagues investigated the biochemical contribution to stochastic gene expression using *Bacillus subtilis* as a model organism (Ozbudak et al., 2002). The authors fused the *gfp* gene in front of an IPTG-inducible promoter and quantified the noise level by FACS analysis. By adding various amounts of IPTG and making targeted mutagenesis in either the promoter region or the ribosomal binding site, the authors were able to determine the source of the noise. In one set of experiments, the transcriptional rate was changed by varying the IPTG concentration and the authors found that the transcriptional efficiency did not significantly affect noise strength. In contrast, by making point mutations in the ribosomal binding site, the authors found a strong positive correlation with translational efficiency (Ozbudak et al., 2002). This study provides the engineer with tools to modify the level of noise for a single gene. Increasing the rate of translation, for instance, increases the noise strength and, conversely, decreasing the rate of translation leads to a reduction in noise strength. If, for example, the desired outcome in protein production is to stay constant, changing the rate of translation can be counterbalanced by changing the rate of transcription. In the model organism *Escherichia coli*, it has been observed that key regulatory proteins display reduced translational rates, thereby minimizing noise at the protein level (the ultimate measure of gene expression) (Ozbudak et al., 2002; Raser & O'Shea, 2005).

In a recent study, Mutalik, Guimaraes, Cambray, Lam, et al. (2013) showed that the identity of variation in genetic elements is more complex. By constructing a full combination library of different promoters and 5′ untranslated regions (UTR) fused to either *gfp* or *rfp*, the researchers monitored the amounts of mRNA and protein for all combination. This systematic approach revealed that the 5′ UTR containing the Shine–Dalgarno element is a key contributing factor to variance (Mutalik, Guimaraes, Cambray, Mai, et al., 2013). Control of the way genetic elements are structured allow a reliable, establishment of basic principles for genetic designs (Mutalik, Guimaraes, Cambray, Lam, et al., 2013).

An important factor contributing to noise is the so-called finite number effect. Basically, the hypothesis predicts that stochastic effects and noise are more prominent when only a few molecules of a process are present. Thus, increasing the number of molecules for a chemical reaction is an effective way of reducing noise (Smits et al., 2006).

Transcriptional bursting, discussed earlier, may not necessarily result in an overall heterogeneous protein pool within a population, since its effects can be buffered by reducing the rate of protein degradation (Raj, Peskin, Tranchina, Vargas, & Tyagi, 2006). Therefore, when constructing a gene regulatory circuit, system noise can be reduced at the protein level by slow decay rates.

Recently, it was suggested that one source of noise that is important for phenotypic variation is transcription fidelity (Gordon et al., 2009). It was found that *E. coli* cells that supposedly display reduced fidelity of RNA polymerase (RNAP) activity, thereby increasing the mistakes in transcribing DNA into RNA, had a perturbed switching frequency of an artificial bistable switch (Gordon et al., 2009). This implies that bistable switches can be used to identify factors that alter processivity and/or fidelity of transcription *in vivo*. For instance, if the transcription of a gene encoding a nonabundant transcriptional regulator is more frequently paused, resulting in the production of less regulator protein, this would have a significant impact on the fraction of cells displaying the phenotype controlled by this regulator. Alternatively, if RNAP pauses less frequently and is more processive, more regulator protein will be produced.

A powerful way to control the level of noise in gene expression is through feedback mechanisms. For instance, it has been proposed and shown experimentally that negative feedback loops (autoregulation) provide robustness and stability in gene networks, thereby reducing noise (Becskei & Serrano, 2000). Using *E. coli* as a model organism, by measuring the coefficient of variance, these authors showed that the degree of variability for an autoregulated system is lower compared to an unregulated system (Becskei & Serrano, 2000). Conversely, positive feedback is known to increase noise (Kaern et al., 2003) and positive feedback (or double-negative feedback) is a main driving force for creating bistable switches, that is, the existence of two stable expression states (Ferrell, 2002; Smits et al., 2006; Ghosh, Banerjee, & Bose, 2012).

5 NOISE AND HETEROGENEITY IN GENE EXPRESSION

At first sight, it might appear counterintuitive that the regulatory systems of important bacterial phenotypes, such as sporulation, biofilm formation and even virulence, rely on stochasticity (Veening, Smits, & Kuipers, 2008; Eldar & Elowitz, 2010). However, for microbes to respond to changes in their microenvironment, they need features like noise in gene expression and as such noisy gene expression driving certain traits might have been selected for during evolution as part of bet-hedging or division of labour strategies. To demonstrate the power of noise-regulated gene circuits, we take a more detailed look at the heterogeneous expression of certain virulence factors.

In order to invade a host, a pathogen needs to overcome many hurdles, like travelling from one part of the body to another through varyingly hostile environments. Resilience to these changing environments requires rapid adaptation. Pathogens use numerous strategies to keep one step ahead of their host, such as the exchange of DNA or a high mutation rate, enabling the pathogen to acquire advantageous traits. However, when a pathogen needs to adapt rapidly, mutation is not efficient enough and the pathogen needs a more sophisticated means of adaptation of the population, for example in the form of division of labour. Pathogens appear to be able to accomplish this by using noise-regulated or noisy gene circuits.

Because of the complex nature of a pathogen's natural environment, from the many interspecies interactions to the number of changing environments a pathogen sometimes must go through, this is a relatively underresearched field. However, for an increasing number of pathogens, ranging from well-known bacteria like *E. coli* to the malaria parasite *P. falciparum*, the last few years have seen an increase in identification of noise-regulated gene circuits that play a role in virulence and pathogenesis (Table 6.1; Butala et al., 2012; Rovira-Graells et al., 2012).

In the following section, we discuss how two pathogenic bacteria employ heterogeneous gene expression when invading a host: *Salmonella enterica* serovar Typhimurium, (short: *Salmonella* Typhimurium) and *Streptococcus pneumoniae*.

6 BISTABLE EXPRESSION OF PNEUMOCOCCAL PILI

It is not well understood why, in most instances, *S. pneumoniae* lives as a harmless commensal organism, but that can suddenly turn into a dangerous pathogen. In fact *S. pneumoniae* is one of the most important human pathogens, responsible for the deaths of nearly 1 million children each year (O'Brien et al., 2009). The consensus is that invasion of a host by *S. pneumoniae* starts with nasopharynx colonization. Disease occurs when *S. pneumoniae* is able to travel to otherwise sterile parts of the body, such as the bloodstream, the inner ear or the lungs. Important virulence factors and pathogenicity islands are necessary for colonization of the host (Weiser, 2010).

One important virulence factor is the type 1 pilus, which *S. pneumoniae* can use for adherence to the surface of the upper respiratory tract during colonization. The reports about the role of pili in virulence are somewhat ambiguous. In murine models, pili have been reported to be important for virulence, while in humans, they do not appear to be associated with increased virulence (Basset et al., 2007). However, after the introduction in 2000 of a vaccine targeting the pilus, there was a rapid decrease of strains with pilus genes isolated in hospitals. At the present time, the percentage of strains able to form pili is back to its old value, around 25%, suggesting that pili can be advantageous for colonization in humans (Regev-Yochay et al., 2010).

Table 6.1 Noise-Regulated/Heterogeneously Expressed (Black) and Possible Noise-Regulated (Grey) Virulence Factors and Traits Contributing to Pathogenicity in Bacteria and Other Pathogens

Organism	System	Reference
Escherichia coli	Production of bacteriotoxin colisin	Butala et al. (2012)
Mycobacterium tuberculosis	Persistence and antibiotic resistance	Wakamoto et al. (2013)
Pseudomonas aeruginosa	*Responsiveness to quorum-sensing signals*	Kohler, Buckling, and van Delden (2009)
	Metabolic state and antibiotic resistance in biofilms	Williamson et al. (2012)
Salmonella enterica serovar Typhimurium	Type 3 secretion system	Ackermann et al. (2008) and Diard et al. (2013)
	Flagellar expression	Cummings, Barrett, Wilkerson, Fellnerova, and Cookson (2005) and Stewart and Cookson (2012)
Streptococcus pneumoniae	Type 1 pilus	Basset et al. (2012) and De Angelis et al. (2011)
	Production of pneumolysin	Ogunniyi, Grabowicz, Briles, Cook, and Paton (2007)
	Expression of capsule	Lysenko, Lijek, Brown, and Weiser (2010)
Vibrio cholerae	Expression of the toxin-coregulated pilus (TCP)	Nielsen et al. (2010)
Candida glabrata (fungi)	Production of adhesin Epa1	Halliwell, Smith, Muston, Holland, and Avery (2012)
Candida albicans (fungi)	Expression of transcription regulator Efg1	Pierce and Kumamoto (2012)
Plasmodium falciparum (protozoa)	Expression of genes involved in host–parasite interactions	Rovira-Graells et al. (2012)

As an additional layer of complication, it has recently been shown that the genes encoding pili formation are heterogeneously expressed (Basset et al., 2011; De Angelis et al., 2011). Pili formation is encoded by the *rlrA* pathogenicity island, containing the regulator RlrA, three cell-anchored surface proteins and three sortases. The surface proteins, RrgA, B and C, contain C-terminal sorting terminals, suggesting that the sortases on the pathogenicity island could be used for this purpose. Not much is known about the regulation of the genes on this island, but it has been shown that the bistability of pilus expression can be altered by changing the expression of RlrA (Basset et al., 2012). RlrA activates expression of the whole pathogenicity island, including itself, but is repressed by one of the surface proteins,

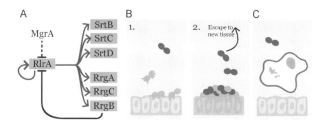

FIGURE 6.2

(A) Pathogenicity island for pili formation is mostly self-regulated. RlrA regulates itself and all six other genes (blue). RrgB inhibits RlrA at the protein level. MgrA, a regulator outside the pathogenicity island, is known to either directly or indirectly repress RlrA activity. (B) Hypothetical role for pili expression in early colonization. Pili-expressing cells (yellow, 1) can attach to the nasopharynx epithelium, stimulating biofilm formation (2). When there is a chance, nonexpressing cells (red) can invade other parts of the body. (C) When the host is primed with RrgB, macrophages recognize the pili-expressing cells, and *S. pneumoniae* will not be able to colonize the nasopharynx. (See color plate.)

RrgB (Figure 6.2A). Only one repressor outside the island has been identified up until now: MgrA (Hemsley, Joyce, Hava, Kawale, & Camilli, 2003). How bistability in pilus formation is established at the molecular level is currently not known, but it is tempting to speculate that noise in gene expression activates the RlrA autostimulatory loop in a stochastic manner and that the combined action of noise and positive feedback is essential in setting up the observed phenotypic variation. Pili on individual cells might be formed by pulses of gene expression in which the RlrA autostimulatory loop is responsible for rapid synthesis of the pilus and the MgrA repressor acts to dampen and switch off gene expression similar to competence development in *B. subtilis*, which is governed by such a noisy excitable bistable switch (Süel, Garcia-Ojalvo, Liberman, & Elowitz, 2006).

Recently, the pilus protein RgrB has been tested as a vaccine target in murine models, showing that this vaccine does protect against *S. pneumoniae* heterogeneously expressing pili. However, not much is known about the function of heterogeneous expression *in vivo*, nor is it known if there is any difference in expression patterns during the different stages of infection. One possible explanation for the effectiveness of the vaccine is that the pili are always expressed at a certain ratio and that this ratio is maintained actively, which, in immunized mice, eventually leads to the eradication of the pathogen. However, Moschioni and colleagues noted that it could also be the case that pilus expression is more highly expressed at early stages of infection (see Figure 6.2B) (Moschioni et al., 2012). This supports the hypothesis that pilus-expressing cells could initiate colonization by adhesion to the epithelium. When the host is primed against pilus-expressing cells, the colonization would be inhibited, even though the pilus-expressing phenotype would only be important at this very early stage (Figure 6.2C).

7 COOPERATIVE VIRULENCE IN *SALMONELLA ENTERICA* S. TYPHIMURIUM

Salmonella Typhimurium is the number one cause of food poisoning in Western countries, causing around one million cases of illnesses in the United States every year. The pathogen is shown to be remarkably adaptive, being able to invade a large range of host organisms, and, within the host, has to go through numerous different environments.

Salmonella Typhimurium invades the host through the Peyer's patches, aggregations of lymphoid tissue in the lowest part of the small intestine. During invasion, *Salmonella* secretes flagellin through the type 3 secretion system (T3SS), which helps to outcompete the natural commensals living both in the Peyer's patches and in the small intestine and evoke an inflammatory response. When the pathogen is taken up by phagocytes, it remains viable and is transported to systemic tissue. Interestingly, in murine models, *Salmonella* Typhimurium isolates show heterogeneous expression of the genes involved in flagella formation. The distribution of this heterogeneity is strikingly different depending on the location of the pathogen. In mouth infections, 100% of the *Salmonella* Typhimurium population is flagellated, but in the Peyer's patches, only part of the population is (Stewart, Cummings, Johnson, Berezow, & Cookson, 2011). Finally, in systemic tissue, flagella formation is repressed.

Heterogeneity of flagellar expression is known to be regulated by the interplay between the flagella master regulator complex $FlhD_4C_2$, its antagonist YdiV and the regulator FliZ (Saini et al., 2010; Stewart et al., 2011; Wada, Tanabe, & Kutsukake, 2011; Moest & Meresse, 2013). A simplified representation of the regulatory circuit is shown in Figure 6.3A (green). $FlhD_4C_2$ activates the expression of the *fliAZ* operon encoding FliZ and sigma factor 28 (σ^{28} or FliA), which in turn regulates the downstream expression of flagellar genes. FliZ stimulates its own (and thereby σ^{28}) expression and simultaneously represses expression of *ydiV*, the product of which blocks the function of $FlhD_4C_2$. Together, this results in a combination of a double-negative feedback and self-stimulation, which meets the characteristics of a noise-sensitive gene circuit (Smits et al., 2006).

While the importance of YdiV and FliZ for heterogeneity is extensively researched *in vitro*, Stewart and colleagues showed that a *ydiV* knockout resulted in a fully flagellated population *in vitro* and in systemic sites in mice (Stewart et al., 2011). In the spleen of mice models, however, *ydiV* mutants and wild-type cells are both unflagellated. Moreover, it has been shown that $FlhD_4C_2$ activity is repressed by degradation by the ClpXP protease, influencing heterogeneity as well (Cummings, Wilkerson, Bergsbaken, & Cookson, 2006; Kage, Takaya, Ohya, & Yamamoto, 2008). ClpXP-knockout mutants are hyperflagellated and, interestingly, their virulence is attenuated. Infection with ClpXP-knockouts protects mice from infection with fully virulent *Salmonella* Typhimurium (Cummings et al., 2006).

These findings, together with the identification of the unique expression pattern of the flagella in the spleen, lead to the hypothesis that flagella are important for the initiation of invasion but that an unflagellated population is needed for perseverance

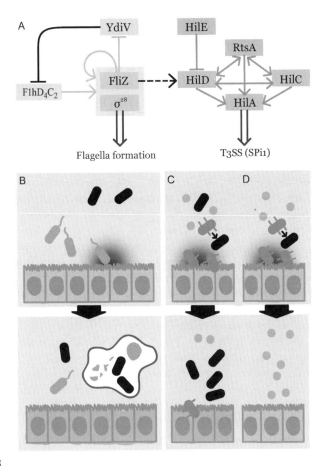

FIGURE 6.3

Role of heterogeneous gene expression during *Salmonella* Typhimurium invasion. (A) Simplified representation of regulation of heterogeneous gene expression of the flagellar proteins (green) and type 3 secretion system (red). Coloured arrows represent gene regulation, black arrows protein–protein interaction. Dashed arrow: not defined whether direct or indirect interaction. The regulatory complex FlhD4C2 activates operon FliAZ containing sigma factor 28 and regulatory protein FliZ. FliZ determines heterogeneous expression of sigma 28 and thereby the flagella formation together with YdiV, which can bind to FldDC. FliZ is also one of the important activators of *hilD* expression. HilD activates *hilA* expression together with RtsA and HilC, which all activate each other and themselves (not depicted for clarity), but HilD is the only essential regulator and thought to influence heterogeneity. (B) Green panel: at early invasion, a flagellated subpopulation reaches the epithelium cells early and triggers the host's immune response. The nonflagellated population can, in contrast to the flagellated population, survive in macrophages and are thereby transported to systemic tissue. (C) Red panel: the majority of T3SS-expressing cells (orange) invade the host epithelium. When a T3SS mutation (black) occurs, the faster-growing nonvirulent subpopulation and the mutants both benefit from the invasion, but a small fraction of the virulent subpopulation remains. However, when a mutation occurs in a 100% virulent population (D), the slow-growing cells will be outcompeted by the mutants who are unable to withstand the commensal population (blue). (See color plate.)

(Figure 6.3B, green). Flagellated organisms have an advantage over unflagellated ones to reach the Peyer's patches through the GI tract. Only a few *Salmonella* Typhimurium cells are sufficient for invading the Peyer's patches, where they reproduce and form a heterogeneous population. Once there, the flagellated population triggers the immune response. The unflagellated population is taken up by macrophages just as well as the flagellated population, but it will not be recognized and can thereby travel to systemic sites.

This cooperation model of the role of flagella during a *Salmonella* Typhimurium invasion cannot be reviewed without looking at the formation of the T3SS, the genes for which are also expressed heterogeneously. Even though the regulation of both systems is connected by two important regulators (FliZ and HilD; see Figure 6.3A), heterogeneity of T3SS expression can be regulated separately from flagellar expression. This provokes the thought that several phenotypes of *Salmonella* Typhimurium might exist during invasion.

The molecular mechanism of T3SS regulation and formation is not fully understood, but expression is controlled by the regulators HilD, RtsA and HilC, which activate their own expression and that of each other, and, importantly that of HilA, the key regulator of the pathogenicity island 1 on which the T3SS system is located. However, HilD is the only essential regulator and is important for heterogeneity in T3SS expression. Its gene and activity is regulated by many proteins, of which FliZ is one of the most important by promoting HilD activity at the protein level (Chubiz, Golubeva, Lin, Miller, & Slauch, 2010).

How the phenotypes resulting from these different gene circuits interact is not known. Stewart and Cookson proposed a model where three distinct populations could be found in a host organism (Stewart & Cookson, 2012). Where the flagella and T3SS would mostly be expressed together, they see a distinct role for flagella-only expression in early invasion, when motility is an important trait for colonization but expression of T3SS evokes inflammatory responses.

CONCLUDING REMARKS

The use of noise-regulated gene circuits and bistable gene expression is widespread among bacteria and appears to be important for pathogenesis: a growing number of reports show a range of pathogenic organisms utilize noisy gene circuits and/or show bistable gene expression (Table 6.1). Importantly, noise-driven phenotypic variation can set up a small subpopulation of cells, which are already prepared for a specific change in the environment in the future such as the presence of a new host. Most of the time, these noise-driven phenotypes can also be activated in nonactivated cells by the 'standard' way, for instance, via sensing mechanisms. These strategies are mostly related to fast adaptation to the different environments and threats inside the host. While mutation, as a means of adaptation, is rigorous and unregulated, noise-mediated gene expression is both fast and reversible.

References

Ackermann, M., Stecher, B., Freed, N. E., Songhet, P., Hardt, W. D., & Doebeli, M. (2008). Self-destructive cooperation mediated by phenotypic noise. *Nature*, *454*, 987–990.

Armitage, B. A. (2011). Imaging of RNA in live cells. *Current Opinion in Chemical Biology*, *15*, 806–812.

Basset, A., Trzcinski, K., Hermos, C., O'Brien, K. L., Reid, R., Santosham, M., et al. (2007). Association of the pneumococcal pilus with certain capsular serotypes but not with increased virulence. *Journal of Clinical Microbiology*, *45*, 1684–1689.

Basset, A., Turner, K. H., Boush, E., Sayeed, S., Dove, S. L., & Malley, R. (2011). Expression of the type 1 pneumococcal pilus is bistable and negatively regulated by the structural component RrgA. *Infection and Immunity*, *79*, 2974–2983.

Basset, A., Turner, K. H., Boush, E., Sayeed, S., Dove, S. L., & Malley, R. (2012). An epigenetic switch mediates bistable expression of the type 1 pilus genes in *Streptococcus pneumoniae*. *Journal of Bacteriology*, *194*, 1088–1091.

Becskei, A., & Serrano, L. (2000). Engineering stability in gene networks by autoregulation. *Nature*, *405*, 590–593.

Blake, W. J., Balazsi, G., Kohanski, M. A., Isaacs, F. J., Murphy, K. F., Kuang, Y., et al. (2006). Phenotypic consequences of promoter-mediated transcriptional noise. *Molecular Cell*, *24*, 853–865.

Butala, M., Sonjak, S., Kamensek, S., Hodoscek, M., Browning, D. F., Zgur-Bertok, D., et al. (2012). Double locking of an *Escherichia coli* promoter by two repressors prevents premature colicin expression and cell lysis. *Molecular Microbiology*, *86*, 129–139.

Chubb, J. R., Trcek, T., Shenoy, S. M., & Singer, R. H. (2006). Transcriptional pulsing of a developmental gene. *Current Biology*, *16*, 1018–1025.

Chubiz, J. E., Golubeva, Y. A., Lin, D., Miller, L. D., & Slauch, J. M. (2010). FliZ regulates expression of the Salmonella pathogenicity island 1 invasion locus by controlling HilD protein activity in *Salmonella enterica* serovar Typhimurium. *Journal of Bacteriology*, *192*, 6261–6270.

Cummings, L. A., Barrett, S. L., Wilkerson, W. D., Fellnerova, I., & Cookson, B. T. (2005). FliC-specific CD4+ T cell responses are restricted by bacterial regulation of antigen expression. *Journal of Immunology*, *174*, 7929–7938.

Cummings, L. A., Wilkerson, W. D., Bergsbaken, T., & Cookson, B. T. (2006). In vivo, fliC expression by *Salmonella enterica* serovar Typhimurium is heterogeneous, regulated by ClpX, and anatomically restricted. *Molecular Microbiology*, *61*, 795–809.

De Angelis, G., Moschioni, M., Muzzi, A., Pezzicoli, A., Censini, S., Delany, I., et al. (2011). The *Streptococcus pneumoniae* pilus-1 displays a biphasic expression pattern. *PLoS One*, *6*, e21269.

de Jong, I. G., Beilharz, K., Kuipers, O. P., & Veening, J. W. (2011). Live cell imaging of *Bacillus subtilis* and *Streptococcus pneumoniae* using automated time-lapse microscopy. *Journal of Visualized Experiments*, *53*, pii: 3145.

de Lorenzo, V., & Perez-Martin, J. (1996). Regulatory noise in prokaryotic promoters: How bacteria learn to respond to novel environmental signals. *Molecular Microbiology*, *19*, 1177–1184.

Diard, M., Garcia, V., Maier, L., Remus-Emsermann, M. N., Regoes, R. R., Ackermann, M., et al. (2013). Stabilization of cooperative virulence by the expression of an avirulent phenotype. *Nature*, *494*, 353–356.

Dublanche, Y., Michalodimitrakis, K., Kummerer, N., Foglierini, M., & Serrano, L. (2006). Noise in transcription negative feedback loops: Simulation and experimental analysis. *Molecular Systems Biology*, *2*, 41.

Dubnau, D., & Losick, R. (2006). Bistability in bacteria. *Molecular Microbiology*, *61*, 564–572.

Eldar, A., & Elowitz, M. B. (2010). Functional roles for noise in genetic circuits. *Nature*, *467*, 167–173.

Elowitz, M. B., Levine, A. J., Siggia, E. D., & Swain, P. S. (2002). Stochastic gene expression in a single cell. *Science*, *297*, 1183–1186.

Femino, A. M., Fay, F. S., Fogarty, K., & Singer, R. H. (1998). Visualization of single RNA transcripts in situ. *Science*, *280*, 585–590.

Ferguson, M. L., Le Coq, D., Jules, M., Aymerich, S., Radulescu, O., Declerck, N., et al. (2012). Reconciling molecular regulatory mechanisms with noise patterns of bacterial metabolic promoters in induced and repressed states. *Proceedings of the National Academy of Sciences of the United States of America*, *109*, 155–160.

Ferrell, J. E. Jr., (2002). Self-perpetuating states in signal transduction: Positive feedback, double-negative feedback and bistability. *Current Opinion in Cell Biology*, *14*, 140–148.

Fraser, H. B., Hirsh, A. E., Giaever, G., Kumm, J., & Eisen, M. B. (2004). Noise minimization in eukaryotic gene expression. *PLoS Biology*, *2*, e137.

Ghosh, S., Banerjee, S., & Bose, I. (2012). Emergent bistability: Effects of additive and multiplicative noise. *The European Physical Journal. E, Soft Matter*, *35*, 11.

Golding, I., Paulsson, J., Zawilski, S. M., & Cox, E. C. (2005). Real-time kinetics of gene activity in individual bacteria. *Cell*, *123*, 1025–1036.

Gordon, A. J., Halliday, J. A., Blankschien, M. D., Burns, P. A., Yatagai, F., & Herman, C. (2009). Transcriptional infidelity promotes heritable phenotypic change in a bistable gene network. *PLoS Biology*, *7*, e44.

Halliwell, S. C., Smith, M. C., Muston, P., Holland, S. L., & Avery, S. V. (2012). Heterogeneous expression of the virulence-related adhesin Epa1 between individual cells and strains of the pathogen *Candida glabrata*. *Eukaryotic Cell*, *11*, 141–150.

Hemsley, C., Joyce, E., Hava, D. L., Kawale, A., & Camilli, A. (2003). MgrA, an orthologue of Mga, Acts as a transcriptional repressor of the genes within the *rlrA* pathogenicity islet in *Streptococcus pneumoniae*. *Journal of Bacteriology*, *185*, 6640–6647.

Kaern, M., Blake, W. J., & Collins, J. J. (2003). The engineering of gene regulatory networks. *Annual Review of Biomedical Engineering*, *5*, 179–206.

Kaern, M., Elston, T. C., Blake, W. J., & Collins, J. J. (2005). Stochasticity in gene expression: From theories to phenotypes. *Nature Reviews Genetics*, *6*, 451–464.

Kage, H., Takaya, A., Ohya, M., & Yamamoto, T. (2008). Coordinated regulation of expression of Salmonella pathogenicity island 1 and flagellar type III secretion systems by ATP-dependent ClpXP protease. *Journal of Bacteriology*, *190*, 2470–2478.

Kaufmann, B. B., & van Oudenaarden, A. (2007). Stochastic gene expression: From single molecules to the proteome. *Current Opinion in Genetics & Development*, *17*, 107–112.

Kohler, T., Buckling, A., & van Delden, C. (2009). Cooperation and virulence of clinical *Pseudomonas aeruginosa* populations. *Proceedings of the National Academy of Sciences of the United States of America*, *106*, 6339–6344.

Larson, D. R., Singer, R. H., & Zenklusen, D. (2009). A single molecule view of gene expression. *Trends in Cell Biology*, *19*, 630–637.

Li, G. W., & Xie, X. S. (2011). Central dogma at the single-molecule level in living cells. *Nature*, *475*, 308–315.

Locke, J. C., & Elowitz, M. B. (2009). Using movies to analyse gene circuit dynamics in single cells. *Nature Reviews Microbiology*, *7*, 383–392.

Lysenko, E. S., Lijek, R. S., Brown, S. P., & Weiser, J. N. (2010). Within-host competition drives selection for the capsule virulence determinant of *Streptococcus pneumoniae*. *Current Biology*, *20*, 1222–1226.

Moest, T. P., & Meresse, S. (2013). Salmonella T3SSs: Successful mission of the secret(ion) agents. *Current Opinion in Microbiology*, *16*, 38–44.

Moschioni, M., De, A. G., Harfouche, C., Bizzarri, E., Filippini, S., Mori, E., et al. (2012). Immunization with the RrgB321 fusion protein protects mice against both high and low pilus-expressing *Streptococcus pneumoniae* populations. *Vaccine*, *30*, 1349–1356.

Munsky, B., Neuert, G., & van Oudenaarden, A. (2012). Using gene expression noise to understand gene regulation. *Science*, *336*, 183–187.

Mutalik, V. K., Guimaraes, J. C., Cambray, G., Lam, C., Christoffersen, M. J., Mai, Q. A., et al. (2013). Precise and reliable gene expression via standard transcription and translation initiation elements. *Nature Methods*, *10*, 354–360.

Mutalik, V. K., Guimaraes, J. C., Cambray, G., Mai, Q. A., Christoffersen, M. J., Martin, L., et al. (2013). Quantitative estimation of activity and quality for collections of functional genetic elements. *Nature Methods*, *10*, 347–353.

Nevozhay, D., Adams, R. M., Van Itallie, E., Bennett, M. R., & Balazsi, G. (2012). Mapping the environmental fitness landscape of a synthetic gene circuit. *PLoS Computational Biology*, *8*, e1002480.

Nielsen, A. T., Dolganov, N. A., Rasmussen, T., Otto, G., Miller, M. C., Felt, S. A., et al. (2010). A bistable switch and anatomical site control *Vibrio cholerae* virulence gene expression in the intestine. *PLoS Pathogens*, *6*, e1001102.

O'Brien, K. L., Wolfson, L. J., Watt, J. P., Henkle, E., Deloria-Knoll, M., McCall, N., et al. (2009). Burden of disease caused by *Streptococcus pneumoniae* in children younger than 5 years: Global estimates. *Lancet*, *374*, 893–902.

Ogunniyi, A. D., Grabowicz, M., Briles, D. E., Cook, J., & Paton, J. C. (2007). Development of a vaccine against invasive pneumococcal disease based on combinations of virulence proteins of *Streptococcus pneumoniae*. *Infection and Immunity*, *75*, 350–357.

Ozbudak, E. M., Thattai, M., Kurtser, I., Grossman, A. D., & van Oudenaarden, A. (2002). Regulation of noise in the expression of a single gene. *Nature Genetics*, *31*, 69–73.

Paige, J. S., Wu, K. Y., & Jaffrey, S. R. (2011). RNA mimics of green fluorescent protein. *Science*, *333*, 642–646.

Pierce, J. V., & Kumamoto, C. A. (2012). Variation in *Candida albicans* EFG1 expression enables host-dependent changes in colonizing fungal populations. *MBio.*, *3*(4), e00117–12. http://dx.doi.org/10.1128/mBio.00117-12.

Querido, E., & Chartrand, P. (2008). Using fluorescent proteins to study mRNA trafficking in living cells. *Methods in Cell Biology*, *85*, 273–292.

Raj, A., Peskin, C. S., Tranchina, D., Vargas, D. Y., & Tyagi, S. (2006). Stochastic mRNA synthesis in mammalian cells. *PLoS Biology*, *4*, e309.

Raj, A., & van Oudenaarden, A. (2008). Nature, nurture, or chance: Stochastic gene expression and its consequences. *Cell*, *135*, 216–226.

Raser, J. M., & O'Shea, E. K. (2005). Noise in gene expression: Origins, consequences, and control. *Science*, *309*, 2010–2013.

Regev-Yochay, G., Hanage, W. P., Trzcinski, K., Rifas-Shiman, S. L., Lee, G., Bessolo, A., et al. (2010). Re-emergence of the type 1 pilus among *Streptococcus pneumoniae* isolates in Massachusetts, USA. *Vaccine, 28*, 4842–4846.

Rovira-Graells, N., Gupta, A. P., Planet, E., Crowley, V. M., Mok, S., Ribas de, P., et al. (2012). Transcriptional variation in the malaria parasite Plasmodium falciparum. *Genome Research, 22*, 925–938.

Saini, S., Koirala, S., Floess, E., Mears, P. J., Chemla, Y. R., Golding, I., et al. (2010). FliZ induces a kinetic switch in flagellar gene expression. *Journal of Bacteriology, 192*, 6477–6481.

Schneider, C. A., Rasband, W. S., & Eliceiri, K. W. (2012). NIH Image to ImageJ: 25 years of image analysis. *Nature Methods, 9*, 671–675.

Shepherd, D. P., Li, N., Micheva-Viteva, S. N., Munsky, B., Hong-Geller, E., & Werner, J. H. (2013). Counting small RNA in pathogenic bacteria. *Analytical Chemistry, 85*, 4938–4943.

Sliusarenko, O., Heinritz, J., Emonet, T., & Jacobs-Wagner, C. (2011). High-throughput, subpixel precision analysis of bacterial morphogenesis and intracellular spatio-temporal dynamics. *Molecular Microbiology, 80*, 612–627.

Smits, W. K., Kuipers, O. P., & Veening, J. W. (2006). Phenotypic variation in bacteria: The role of feedback regulation. *Nature Reviews Microbiology, 4*, 259–271.

Stewart, M. K., & Cookson, B. T. (2012). Non-genetic diversity shapes infectious capacity and host resistance. *Trends in Microbiology, 20*, 461–466.

Stewart, M. K., Cummings, L. A., Johnson, M. L., Berezow, A. B., & Cookson, B. T. (2011). Regulation of phenotypic heterogeneity permits Salmonella evasion of the host caspase-1 inflammatory response. *Proceedings of the National Academy of Sciences of the United States of America, 108*, 20742–20747.

Süel, G. M., Garcia-Ojalvo, J., Liberman, L. M., & Elowitz, M. B. (2006). An excitable gene regulatory circuit induces transient cellular differentiation. *Nature, 440*, 545–550.

Swain, P. S., Elowitz, M. B., & Siggia, E. D. (2002). Intrinsic and extrinsic contributions to stochasticity in gene expression. *Proceedings of the National Academy of Sciences of the United States of America, 99*, 12795–12800.

Thattai, M., & van Oudenaarden, A. (2001). Intrinsic noise in gene regulatory networks. *Proceedings of the National Academy of Sciences of the United States of America, 98*, 8614–8619.

Trcek, T., Chao, J. A., Larson, D. R., Park, H. Y., Zenklusen, D., Shenoy, S. M., et al. (2012). Single-mRNA counting using fluorescent in situ hybridization in budding yeast. *Nature Protocols, 7*, 408–419.

Veening, J. W., Smits, W. K., & Kuipers, O. P. (2008). Bistability, epigenetics, and bet-hedging in bacteria. *Annual Review of Microbiology, 62*, 193–210.

Wada, T., Tanabe, Y., & Kutsukake, K. (2011). FliZ acts as a repressor of the *ydiV* gene, which encodes an anti-FlhD4C2 factor of the flagellar regulon in *Salmonella enterica* serovar Typhimurium. *Journal of Bacteriology, 193*, 5191–5198.

Wakamoto, Y., Dhar, N., Chait, R., Schneider, K., Signorino-Gelo, F., Leibler, S., et al. (2013). Dynamic persistence of antibiotic-stressed mycobacteria. *Science, 339*, 91–95.

Weiser, J. N. (2010). The pneumococcus: Why a commensal misbehaves. *Journal of Molecular Medicine (Berlin), 88*, 97–102.

Williamson, K. S., Richards, L. A., Perez-Osorio, A. C., Pitts, B., McInnerney, K., Stewart, P. S., et al. (2012). Heterogeneity in *Pseudomonas aeruginosa* biofilms includes expression of ribosome hibernation factors in the antibiotic-tolerant subpopulation and hypoxia-induced stress response in the metabolically active population. *Journal of Bacteriology, 194*, 2062–2073.

Young, J. W., Locke, J. C., Altinok, A., Rosenfeld, N., Bacarian, T., Swain, P. S., et al. (2012). Measuring single-cell gene expression dynamics in bacteria using fluorescence time-lapse microscopy. *Nature Protocols*, *7*, 80–88.

Yu, J., Xiao, J., Ren, X., Lao, K., & Xie, X. S. (2006). Probing gene expression in live cells, one protein molecule at a time. *Science*, *311*, 1600–1603.

Zenklusen, D., Larson, D. R., & Singer, R. H. (2008). Single-RNA counting reveals alternative modes of gene expression in yeast. *Nature Structural & Molecular Biology*, *15*, 1263–1271.

CHAPTER 7

Platforms for Genetic Design Automation

Chris J. Myers[1]

Department of Electrical and Computer Engineering University of Utah, Salt Lake City, Utah, USA
[1]*Corresponding author. e-mail address: myers@ece.utah.edu*

1 INTRODUCTION

Recently, the new field of *synthetic biology* is generating tremendous interest and excitement due to a wide variety of potential applications from producing cheaper drugs (Ro et al., 2006), to optimizing the production of biofuels (Atsumi & Liao, 2008), and even potentially to treat diseases such as cancer (Anderson, Clarke, & Arkin, 2006). While the field of *genetic engineering* has been around for more than 30 years, synthetic biology adds the concepts of standards for data representation, abstraction at various levels, and a decoupling of the design and construction process (Endy, 2005). These concepts create an engineering discipline that ultimately allows for the development of automated design methods and software tools. Inspiration for this development work can be drawn from the field of electronic design automation (EDA) that has enabled the production of ever more complex integrated circuits each year powering the information age that we live in today. To enable the coming biology age, new genetic design automation (GDA) platforms are required (Myers, 2009; Myers, Barker, Kuwahara, et al., 2009). The subject of this chapter is to give a snapshot of the current state-of-the-art in GDA software tools. It should be emphasized that this is a fast-moving field, so undoubtedly, there are new tools not described here and even the tools described will have many new, exciting features at the time of this publication. However, this chapter should give the potential users of platforms for GDA an idea of the types of support they can expect from these software tools while enabling developers of GDA tools a broader perspective of the field. Finally, it can hopefully begin a dialogue between the users and developers to ensure that the GDA tools being developed are indeed meeting the needs for genetic design.

Figure 7.1 shows a typical genetic circuit design workflow. It begins with a high-level specification describing the desired function for the genetic circuit design. This specification should then be analysed using simulation or other analysis techniques to verify that the specification does indeed meet the design requirements. If it does not meet the requirements, then the specification needs to be revised. While the specification itself does not need to be written using a standard language, if it can be compiled into a standard modelling language, such as the Systems Biology Markup

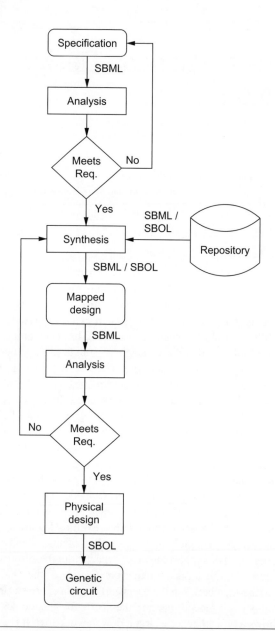

FIGURE 7.1

Genetic circuit design workflow.

Language (SBML) described in the succeeding text, then a wider variety of analysis tools become available. Once the specification has been verified, the next step is to synthesize a mapped genetic circuit design by selecting parts from a repository that together implement the specified behaviour. Ideally, the structure of both the parts and mapped design is represented using the Synthetic Biology Open Language (SBOL). The mapped design should then be analysed to verify that it meets the design requirements, and if it does not, then synthesis may need to be performed again using different parts. Once again, the use of the SBML standard facilitates this step. Once the mapped design has been verified, it is passed to physical design tools to prepare it for construction. These tools may optimize the DNA sequence, produce a plan for assembly, and even control robots to create the genetic circuit in the laboratory.

This chapter is organized as follows. Section 2 introduces the SBML and SBOL standards used by many GDA tools. Section 3 describes the repositories for storing information about genetic circuits. Section 4 presents the capabilities of many GDA tools that are available today. Finally, Section 5 summarizes the state-of-the-art and future directions for GDA platforms.

2 STANDARDS

Unless a GDA tool implements all the steps shown in Figure 7.1, it is necessary for the tool to exchange data with other GDA tools. To facilitate this exchange, a GDA tool can use standard data representations. As mentioned earlier, SBML is the dominating standard for the exchange of biological models, while SBOL is an emerging standard for describing the structure of genetic designs. This section briefly describes these two standards.

2.1 Systems biology markup language

SBML is a modelling standard developed by the systems biology community (Hucka et al., 2003). SBML provides a common intermediate format for biochemical models. The development of the SBML standard has been extremely successful. SBML, for example, is currently supported by more than 250 open-source and commercial tools that allow researchers to create, annotate, simulate, and visualize models. Most importantly, they allow researchers to exchange models and deposit them in the BioModels Database (Chelliah, Laibe, & Novère, 2013). SBML Level 3 Core can represent reaction-based models of biological systems. As shown in Figure 7.2, an SBML model can include various *compartments* (i.e., the cytoplasm (c) and nucleus (n)), *chemical species* (i.e., protein A and protein B) and *chemical reactions* (i.e., the conversion of protein B into protein A with a rate specified by the function $f_1(x)$). SBML models can also include *parameters*, *functions*, *unit definitions*, *initial assignments*, *rules* for continuous relationships, *events* for discontinuous state changes, and *constraints* to indicate when a simulation should terminate. One of the major enhancements of SBML Level 3 over SBML Level 2 is the support for packages to extend the modelling capabilities of SBML beyond its *core* elements. An SBML package can add either completely new elements to a model or simply new parts to existing elements. For example, a layout package may add to a species its x and y coordinates for a

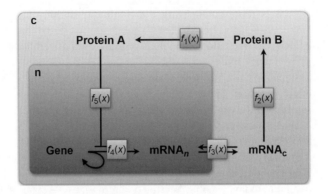

FIGURE 7.2

Basic features of SBML. (See color plate.) Courtesy of Mike Hucka.

graphical editor. Several packages have been completed including *layout*, *hierarchical model composition*, *qualitative modelling*, and *flux balance constraints*.

2.2 Synthetic biology open language

SBOL, being developed by the synthetic biology community, is a standard for describing genetic parts and designs (Galdzicki et al., 2012, 2013). As shown with the unified modelling language (UML) diagram in Figure 7.3, SBOL consists of collections of annotated DNA, RNA, or protein *component* sequences. For example, a DNA component may be a segment of DNA that has a particular function such as a *promoter*, *open reading frame* (i.e., gene), *ribosome binding site*, and *terminator*. The type of a component is indicated using a type from the sequence ontology (SO) (Eilbeck et al., 2005). A component can also be a sequence that is composed of other components hierarchically. Each annotation indicates the location on the start and end point of the annotation within the sequence and also the strand on which it is located in the case of DNA components. The order of annotations can also be simply ordered using the precedes relation when the sequence is not yet known. Currently, SBOL only includes structural information about DNA components, but extending it to RNA and protein components is fairly straightforward. Additional extensions to SBOL currently under consideration are shown with dashed boxes in Figure 7.3. These include the *device* element that is a collection of components that provide a particular function. Devices can be *primitive* (i.e., composed of only other components) or *composite* (i.e., composed of other devices). The *system* element pairs a set of devices with their *experimental context* represented with the *context* element. The context for a system includes information like the strain of the host, the medium in which the host resides, the container in which the medium is stored, the environmental conditions, and the measurement device used to study the system. The *model* element is proposed to connect models to systems. SBOL does not attempt to encode models but rather it refers to models written in

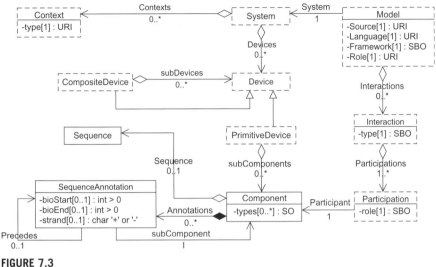

FIGURE 7.3

UML diagram for SBOL.

modelling languages such as SBML. Therefore, the model element includes a reference to the source file for the model, the language that it is written in, the type of the modelling framework used, and the role of the model. Finally, the *interaction* and *participation* elements are proposed to describe relationships between components. For example, a protein component for a transcription factor may repress a DNA component for a promoter. Each interaction has a type from the systems biology ontology (SBO) (Courtot et al., 2011) and a list of *participations*. Each participation refers to a component and the role that component has in the interaction. For the previous example, the protein component's role would be repressor, while the DNA component's role would be repressed.

3 REPOSITORIES

A critical first step to GDA is to collect a set of genetic parts from which designs can be constructed. While at a minimum an annotated DNA sequence is required, a sequence alone is usually not sufficient to reproduce the desired behaviour (Peccoud et al., 2011). Therefore, a part repository ideally should also include additional information about the part including the host strain in which it is to be used and the environment in which the host resides. Furthermore, additional information such as kinetic parameters and perhaps even a model for the part would be extremely useful. This section describes several repositories in which genetic parts and their models can be stored.

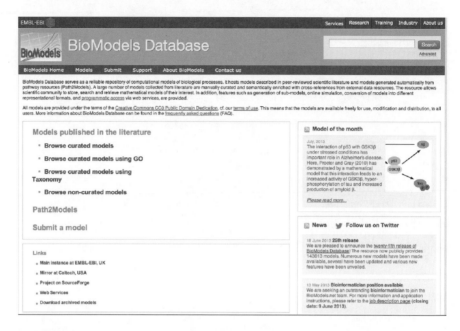

FIGURE 7.4

Screenshot for the BioModels Database. (See color plate.)

3.1 BioModels database

SBML models can be archived and shared using the *BioModels Database* shown in Figure 7.4 (Chelliah et al., 2013). In June of 2013, the 25th release of models was announced that includes 143,013 models. This release includes 963 models published in the literature and 142,050 models generated as part of the Path2Models project. The curated models in the database have been carefully annotated with ontology terms, connections to other repositories describing components of the model, and the publication(s) that describe the model. While most models come from systems biology research, the database does include some synthetic biology designs such as the repressilator (Elowitz & Leibler, 2000). The database provides various ways to search for information, routines to export information about the model in various formats and visualizations, and a web service interface to allow GDA tools to connect to the repository.

3.2 GenBank

A naive approach to store structural information for genetic parts would be to enter them into the GenBank database (Bensen et al., 2013) shown in Figure 7.5. GenBank is the National Institute of Health (NIH) database for annotated gene sequences. GenBank is ideally suited for bioinformatics researchers who are interested in analyzing DNA sequences, but it is not well suited to synthetic biology because it is

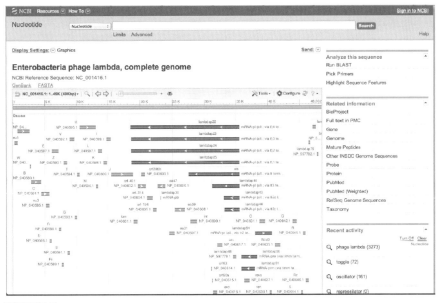

FIGURE 7.5

Screenshot for GenBank. (See color plate.)

incapable of storing additional information required for construction and use of a genetic part. For those researchers though who wish to use GenBank, there is a converter as part of the Joint BioEnergy Institute (JBEI) tools described in the succeeding text that can convert between GenBank's file format and the SBOL standard format enabling sequence information to be exchanged between the GenBank repository and GDA tools.

3.3 Registry of standard biological parts (iGEM registry)

The *Registry of Standard Biological Parts* (or iGEM registry), shown in Figure 7.6, addresses many of these limitations by enabling the inclusion of more information about the use of genetic designs within their repository (see http://parts.igem.org). In addition to an annotated sequence, the iGEM registry includes extensive documentation, such as, detailed descriptions of the part, information about how the part can be used, the quality of the part as assessed by user experiences, and characterization information, when available. The maintainers of this repository also provide the physical DNA for many of the parts to iGEM teams and other researchers. The registry provides a custom XML interface to allow tools to search for and use parts. However, the registry does not currently include the ability to import and export SBOL. To address this limitation, there does exist a converter from the iGEM registry's XML format into the SBOL standard (instructions for its use are available on the SBOL website http://www.sbolstandard.org).

FIGURE 7.6

Screenshot for the iGEM Registry. (For colour version of this figure, the reader is referred to the online version of this chapter.)

3.4 Standard biological parts knowledgebase

The standard biological parts knowledgebase (SBPkb), shown in Figure 7.7, utilizes SBOL and leverages semantic web technology to create a parts repository that can be readily searched using SPARQL queries (Galdzicki, Rodriguez, Chandran, Sauro, & Gennari, 2011). This enables synthetic biologists to more readily find a genetic part with desired properties, and it enables these parts to be easily imported into their GDA tools for further analysis and design steps. Currently, this repository does not include part characterization data and other information required for the modelling of genetic designs.

3.5 BioFab

To address the need for characterization data, the BioFab project was created by the National Science Foundation. The goal of the BioFab is to develop a collection of characterized standard biological parts that are made freely available to the synthetic biology community. Data on these characterized parts are made available by the BioFab using a data access client shown in Figure 7.8. Currently, the database includes data on various promoters and terminators for use in *E. coli* (Cambray et al., 2013; Mutalik, Guimaraes, Cambray, Lam, et al., 2013; Mutalik, Guimaraes, Cambray, Mai, et al., 2013). While valuable, these data are not provided using a standard.

3 Repositories   185

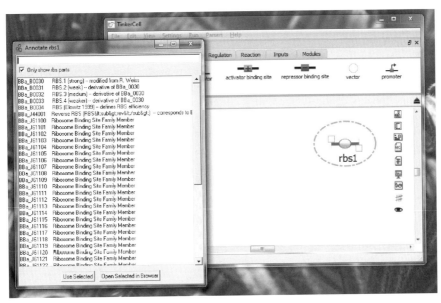

FIGURE 7.7

Screenshot for WikiDust interface to SBPkb. (See color plate.)

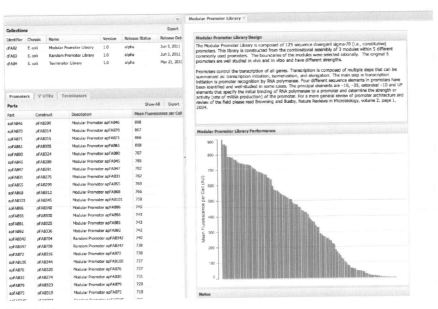

FIGURE 7.8

Screenshot for the BioFab data access client. (For colour version of this figure, the reader is referred to the online version of this chapter.)

3.6 BacilloBricks

To address the need for a standard for the composition and exchange of models of characterized parts, the standard virtual parts (SVPs) formalism is being developed (Cooling et al., 2010). Virtual parts standardize the description of genetic parts such as promoters and ribosome binding sites in a form that includes modular and reusable dynamic models of the functionality introduced by these parts. The *BacilloBricks* repository, shown in Figure 7.9, includes about 3000 SVPs for *Bacillus subtilis*. This repository provides parts in SBOL format and also models of parts in SBML format. SVPs can be composed manually or using computational intelligence based tools such as SVPWrite (Hallinan, Gilfellon, Misirli, Phillips, & Wipat, 2013). The `MoSeC` tool converts appropriately annotated models of synthetic systems into genetic descriptions of the systems (Misirli et al., 2011). The resulting DNA sequences can then be used to guide the assembly or synthesis of DNA for system implementation.

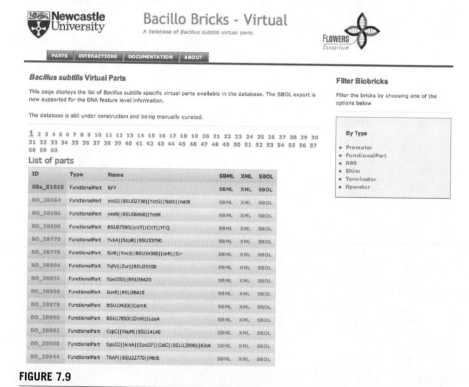

FIGURE 7.9

Screenshot for BacilloBricks. (See color plate.)

FIGURE 7.10

Screenshot for ICE. (See color plate.)

3.7 Inventory of composable elements

The last repository of note is the Inventory of Composable Elements (ICE), which is being developed by the JBEI (Ham et al., 2012). Like the other repositories, ICE allows synthetic biologists to archive both their annotated DNA sequences and other critical information needed to use their part. One unique feature is the integration of the repository with the VectorEditor tool that enables creation and visualization of the DNA sequences. The ICE repository can also be installed by users locally in order to allow the maintenance of a local repository of parts at their own institution. Finally, as shown in Figure 7.10, ICE allows for the import and export of parts described using the SBOL standard (Figure 7.10).

4 GDA SOFTWARE TOOLS

Recently, many GDA software tools have been developed that cover all aspects of the genetic design workflow described earlier. There are GDA tools to create genetic designs at various levels of abstraction from high-level languages down to the DNA sequence. This section briefly describes the capabilities and features of several GDA tools.

4.1 BioJADE

Perhaps one of the first GDA tools is the BioJADE tool shown in Figure 7.11 (Goler, 2004). The BioJADE tool, inspired by EDA tools for integrated VLSI circuits, provides a schematic capture tool for constructing genetic designs, as well as various simulation engines to analyse genetic designs. A key feature of BioJADE is its ability to connect to the Registry of Standard Biological Parts. While a promising GDA tool with many useful features, development of this tool has not continued.

4.2 GenoCAD

A more recent GDA tool with similar capabilities that is under active development at Virginia Tech is GenoCAD (Cai, Wilson, & Peccoud, 2010). As shown in Figure 7.12, GenoCAD also includes a schematic capture tool that allows a user to construct a genetic design from a local library of parts depicted using the SBOL visual standard (Quinn et al., 2013). A unique feature of GenoCAD is that the genetic design is validated with a grammar to ensure that the parts are ordered in a functional manner. After a design is constructed, an ODE model can be produced and simulated with

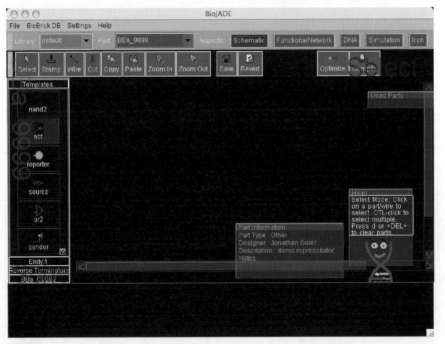

FIGURE 7.11

Screenshot for BioJADE. (See color plate.)

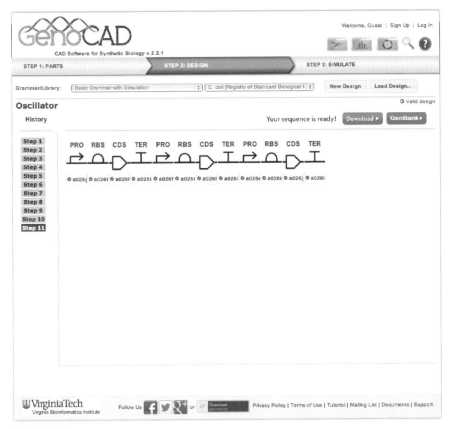

FIGURE 7.12

Screenshot for GenoCAD. (See color plate.)

the COPASI simulator (Hoops et al., 2006). The generated model can also be exported in SBML for analysis with other GDA tools.

4.3 TinkerCell

TinkerCell, developed at the University of Washington, also provides a schematic capture tool that uses SBOL visual symbols and simulation using COPASI (Chandran, Bergmann, & Sauro, 2009). In addition, as shown in Figure 7.13, TinkerCell allows the user to add regulation constructs and biochemical reactions and connect to other analysis procedures within COPASI including stochastic simulation. TinkerCell can import and export SBML models, and it has a connection to the SBPkb repository for genetic parts. Finally, it has excellent support for hierarchical design. TinkerCell, unfortunately, is not currently under active development.

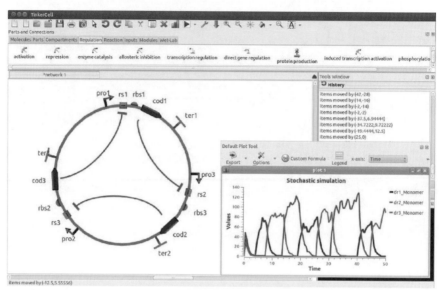

FIGURE 7.13

Screenshot for TinkerCell. (See color plate.)

4.4 Process modelling tool

The process modelling tool (ProMoT) developed at the Max Planck Institute for Dynamics of Complex Technical Systems is yet another tool for the construction of models of genetic designs (Trankle et al., 2000; Marchisio & Stelling, 2008; Mirschel et al., 2009). Models can be described in a textual language, either constructed using a visual editor or imported in SBML. As shown in Figure 7.14, ProMoT models can also be exported in the SBML format for analysis by other tools.

4.5 Synthetic biology software suite

The Synthetic Biology Software Suite (SynBioSS) developed at the University of Minnesota focuses on efficient stochastic simulation of genetic designs to enable rational design (Hill, Tomshine, Weeding, Sotiropoulos, & Kaznessis, 2008). SynBioSS generates models from characterized parts. As shown in Figure 7.15, these models can be analysed within SynBioSS or exported using SBML. The characterization data are obtained from data stored in a database built upon MediaWiki.

4.6 Synthetic biology reusable optimization methodology

The Synthetic Biology Reusable Optimization Methodology (SBROME) developed at the University of California, Davis, begins to address the synthesis problem for genetic circuit designs (Huynh, Tsoukalas, Kppe, & Tagkopoulos, 2013). As shown

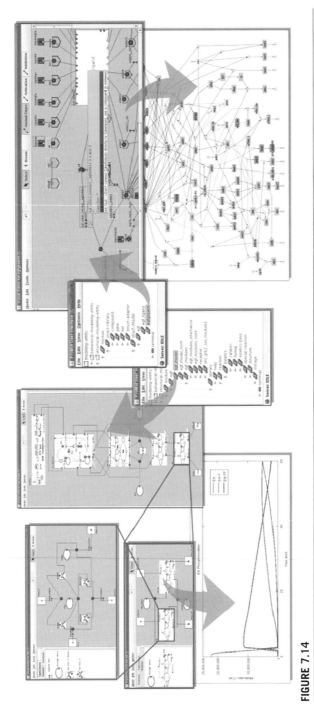

FIGURE 7.14

Screenshot for ProMoT. (See color plate.)

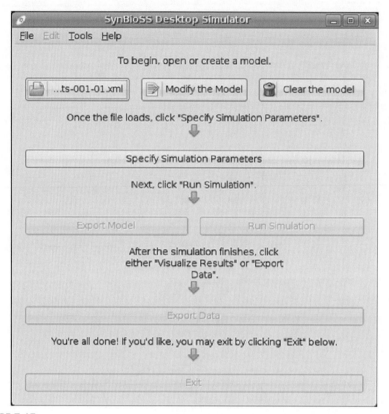

FIGURE 7.15

Screenshot for SynBioSS. (For colour version of this figure, the reader is referred to the online version of this chapter.)

in Figure 7.16, a user can describe a genetic circuit at a higher level of abstraction using Boolean logic gates, and SBROME attempts to select a set of genetic parts from a given library to implement the desired function.

4.7 Genetic engineering of cells

Another GDA tool that performs synthesis is genetic engineering of cells (GEC) being developed at Microsoft Research (Pedersen & Phillips, 2009). As shown in Figure 7.17, a user can describe a genetic circuit structure using a programming language, and GEC determines a set of genetic parts that match that structure. GEC is also capable of generating a model for the design that can be exported in SBML or MATLAB code for further analysis.

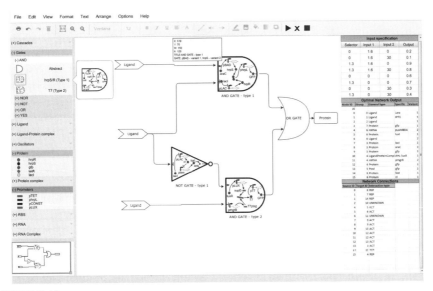

FIGURE 7.16

Screenshot for SBROME. (See color plate.)

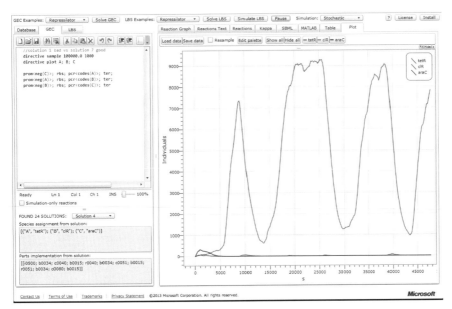

FIGURE 7.17

Screenshot for GEC. (For colour version of this figure, the reader is referred to the online version of this chapter.)

4.8 Kera

Another programming language for synthetic biology is Kera that is being developed at the University of Kerala (Umesh, Naveen, Rao, & Nair, 2010). As shown in Figure 7.18, genetic designs can be created in Kera using either an object-oriented programming environment or a graphical interface that uses SBOL visual symbols. Design validity can be checked using a set of rules, and SBML models can be imported and simulated using Scilab. Finally, Kera includes plug-ins for codon optimization, genome viewing and sequence editing.

4.9 Intelligent biological simulator

Another GDA tool that integrates analysis and synthesis is the Intelligent Biological Simulator (iBioSim) being developed at the University of Utah (Myers, Barker, Jones, et al., 2009; Madsen et al., 2012; Stevens & Myers, 2013). As shown in Figure 7.19, iBioSim includes a schematic capture tool that allows a user to construct a model of the regulatory behaviour of a genetic circuit. Since iBioSim supports all of SBML Level 3, a model can also include biochemical reactions, discrete events, continuous update rules, and desired behaviour specified as constraints. iBioSim supports both the import and export of SBML L3V1 models, and it has a direct

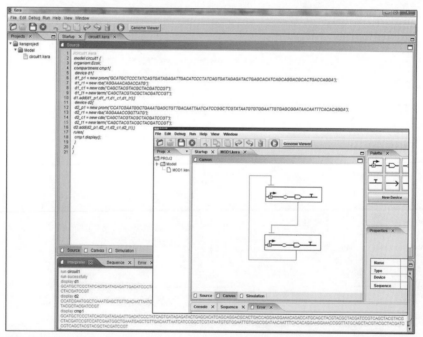

FIGURE 7.18

Screenshot for Kera. (For colour version of this figure, the reader is referred to the online version of this chapter.)

4 GDA Software Tools

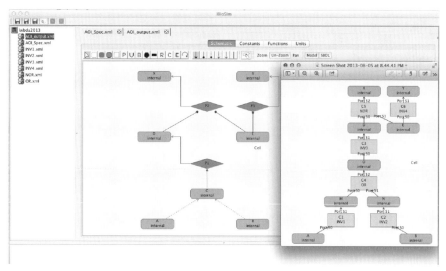

FIGURE 7.19

Screenshot for iBioSim. (See color plate.)

connection to the BioModels Database. `iBioSim` can also import genetic parts expressed in SBOL, annotate SBML elements with these parts, and export the resulting composite genetic design in SBOL. `iBioSim` includes numerous ODE and stochastic simulators. A key feature is the support of reaction-based and logical abstraction enabling efficient analysis techniques such as *stochastic model checking*. As shown in Figure 7.19, support has recently been added to `iBioSim` to synthesize a genetic design by selecting a noninterfering set of parts from a parts library that implements the specified behaviour.

4.10 Sequence editors and optimizers

In addition to `ICE`, JBEI provides several other tools to support genetic design. First, there is the `DeviceEditor`, shown in Figure 7.20, that provides a schematic capture tool to enable the construction of genetic designs using SBOL visual symbols (Chen, Densmore, Ham, Keasling, & Hillson, 2012). Next, there is the `VectorEditor`, shown in Figure 7.21, that provides a schematic capture tool for editing annotated DNA sequences (Ham et al., 2012). Finally, there is `j5` that creates an assembly plan for genetic designs considering a variety of factors including the cost of different alternatives (Hillson, Rosengarten, & Keasling, 2012). Both `VectorEditor` and `j5` can import and export SBOL.

There are also a few commercial GDA tools that provide similar functionality to the JBEI tools including TeselaGen, which is commercializing the JBEI tools themselves. DNA 2.0 produces a tool called `GeneDesigner` that allows a user to edit,

196 CHAPTER 7 Platforms for Genetic Design Automation

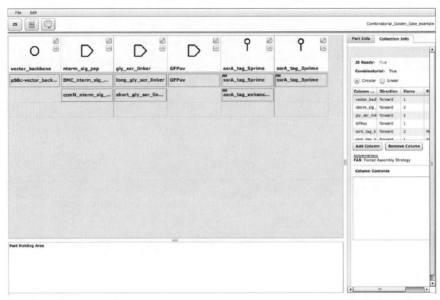

FIGURE 7.20

Screenshot for DeviceEditor. (For colour version of this figure, the reader is referred to the online version of this chapter.)

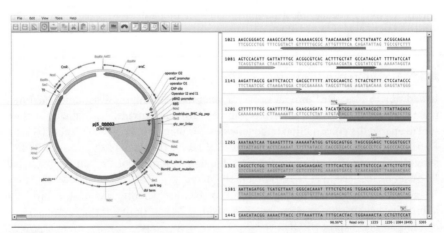

FIGURE 7.21

Screenshot for VectorEditor. (See color plate.)

optimize, and order a DNA sequence for a genetic design. `GenomeCompiler` provided by Genome Compiler provides similar functionality but allows the user to order from multiple vendors.

4.11 Tool chains

There have been a couple efforts to create a complete genetic design workflows by stitching together several independent GDA tools. The first is the Tool-Chain to Accelerate Synthetic Biological Engineering (TASBE) project at BBN Technologies (Beal et al., 2012). The `TASBE` workflow begins with a specification written in the `Proto` high-level language, which is compiled into a dataflow graph (DFG). This DFG is compiled using the `Proto BioCompiler` into an abstract genetic regulatory network (AGRN). The `MatchMaker` tool is then applied to select the concrete genetic parts to implement the AGRN and produce a DNA sequence. Finally, the `Puppeteer/BioCAD` tool is applied to construct an assembly plan and ultimately a physical sample.

The other major tool chain effort is the `Clotho` project at Boston University shown in Figure 7.22 (Xia et al., 2011). A genetic design can be created in `Clotho` either textually using the `Eugene` language (Bilitchenko et al., 2011) or schematically

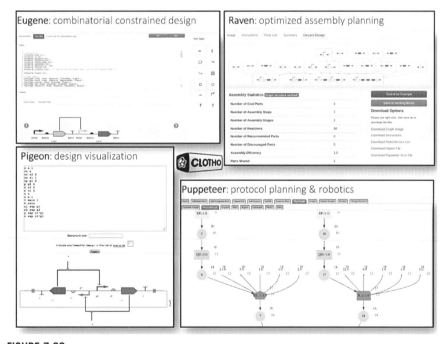

FIGURE 7.22

Screenshots for Clotho. (See color plate.)

using the `Cello` tool. `Eugene` can produce a set of rule-compliant designs from a given set of possible devices, and it is also utilized by JBEI's `DeviceEditor` and `j5` and Life Technologies' `Vector NTI`. The `Cello` and `MatchMaker` tools can be utilized to synthesize a concrete design using genetic parts from either the iGEM Registry or an ICE repository. The design output by `Cello` or `Matchmaker` is passed on to an assembly method planning tool called `Raven`, which optimizes the assembly graph used to assemble the design. Finally, the assembly graph is passed on to `Puppeteer` that generates the protocol instructions for assembling the designs. Information about genetic parts and designs can be exchanged with the `Clotho` tools using SBOL. Finally, designs can be visualized using SBOL visual compliant symbols using the `Pigeon` tool (Bhatia & Densmore, 2013).

5 DISCUSSION

Table 7.1 summarizes the features supported by the GDA tools described in this chapter. As described earlier, the tools support a diversity of features at various levels of abstraction. This fact makes it a bit difficult to develop a perfect taxonomy of feature support. However, this table should give one a rough idea of the capabilities of each GDA tool. The design editor column indicates if a GDA tool has support to create a genetic design using either a schematic editor, a custom language, or both. The analysis column indicates if the GDA tool provides support for stochastic, ODE, logical, or all three types of analysis. The synthesis column indicates whether or not the GDA tool can automatically create a genetic design from a specification using parts from a genetic library. The physical design column indicates whether the tool supports construction aspects such as sequence editing, codon optimization, and assembly planning. The repository connection indicates which model and/or genetic part libraries that the tool can connect to directly. The SBML and SBOL support columns denote whether the tool is capable of importing, exporting, or both these standard file formats for models and genetic components. The platform columns state whether the tool is a web application, Java (i.e., universal) application, or which operating systems are supported. Finally, the active column informs whether the GDA tool is under active development.

One fact is clear from this table which is that no GDA tool supports all aspects of genetic design. This fact makes the support of standards by these tools critically important. For example, if a set of tools support the SBML and SBOL standards, then it is possible to create a specification for a design at a high level of abstraction in one tool, analyse a model of that design in another tool, synthesize a concrete implementation in yet another tool from a repository that provides a library of genetic components using these standards, and perform physical design and assembly planning in a final tool. Therefore, a critical aspect going forward in the development of platforms for GDA will continue to be the improvement of the standards used for data exchange between GDA tools. This effort coupled with the continued progress in the quality of the GDA tools should enable genetic designs to scale and meet the increasing demands of the exciting future applications of synthetic biology.

Table 7.1 Summary of GDA tool features

GDA tool	Design editor	Analysis	Syn.	Phys.	Repository connection	SBML support	SBOL support	Platforms	Act.
BioJADE	Sch.	Stoc	No	No	iGEM	None	None	Java	No
GenoCAD	Sch.	ODE	No	No	None	Export	None	Web	Yes
TinkerCell	Sch.	Stoc/ODE	No	No	SBPkb	Imp/exp	None	All	No
ProMoT	Both	None	No	No	None	Imp/exp	None	Win/lin	Yes
SynBioSS	None	Stoc/ODE	No	No	None	Imp/exp	None	All	Yes
SBROME	Sch.	None	Yes	No	None	None	None	All	Yes
GEC	Lang.	Stoc/ODE	Yes	No	None	Export	None	Web	Yes
Kera	Lang.	ODE	Yes	No	None	Import	None	Java	Yes
iBioSim	Sch.	All	Yes	No	BioModels	Imp/exp	Imp/exp	Java	Yes
JBEI Tools	Sch.	None	No	Yes	ICE	None	Imp/exp	Web	Yes
GeneDesigner	Sch.	None	No	Yes	BioFAB	None	None	Java	Yes
GenomeCompiler	Sch.	None	No	Yes	iGEM	None	None	Win/OSx	Yes
TASBE	Lang.	ODE/log	Yes	Yes	None	None	Export	Web	Yes
Clotho	Both	None	Yes	Yes	iGEM/ICE	None	Imp/exp	All	Yes

Acknowledgements

The author of this work is supported by the National Science Foundation under Grant No. CCF-1218095. Any opinions, findings, and conclusions or recommendations expressed in this material are those of the author(s) and do not necessarily reflect the views of the National Science Foundation. We would also like to thank Nathan Hillson, Michal Galdzicki, Swapnil Bhatia, Marchisio Mario Andrea, Mandy Wilson, Robert Cox, Deepak Chandran, Aaran Adler, Umesh P, Raik Gruenberg, Randy Rettberg, Ilias Tagkopoulos, and Goksel Misirli for feedback on this manuscript.

References

Anderson, J. C., Clarke, E. J., & Arkin, A. P. (2006). Environmentally controlled invasion of cancer cells by engineering bacteria. *Journal of Molecular Biology*, *355*, 619–627.

Atsumi, S., & Liao, J. C. (2008). Metabolic engineering for advanced biofuels production from Escherichia coli. *Current Opinion in Biotechnology*, *19*, 414–419 Tissue, cell and pathway engineering.

Beal, J., Weiss, R., Densmore, D., Adler, A., Appleton, E., Babb, J., et al. (2012). An end-to-end workflow for engineering of biological networks from high-level specifications. *ACS Synthetic Biology*, *1*, 317–331.

Bensen, D. A., Cavanaugh, M., Clark, K., Karsch-Mizrachi, I., Lipman, D. J., Ostell, J., et al. (2013). Genbank. *Nucleic Acids Research*, *41*, 36–42.

Bhatia, S., & Densmore, D. (2013). Pigeon: A design visualizer for synthetic biology. *ACS Synthetic Biology*, *2*, 348–350.

Bilitchenko, L., Liu, A., Cheung, S., Weeding, E., Xia, B., Leguia, M., et al. (2011). Eugene— A domain specific language for specifying and constraining synthetic biological parts, devices, and systems. *PLoS ONE*, *6*(4), e18882.

Cai, Y., Wilson, M. L., & Peccoud, J. (2010). GenoCAD for iGEM: A grammatical approach to the design of standard-compliant constructs. *Nucleic Acids Research*, *38*, 2637–2644.

Cambray, G., Guimaraes, J. C., Mutalik, V. K., Lam, C., Mai, Q.-A., Thimmaiah, T., et al. (2013). Measurement and modeling of intrinsic transcription terminators. *Nucleic Acids Research*, *41*, 5139–5148.

Chandran, D., Bergmann, F. T., & Sauro, H. M. (2009). TinkerCell: Modular CAD tool for synthetic biology. *Journal of Biological Engineering*, *3*(19).

Chelliah, V., Laibe, C., & Novère, N. L. (2013). Biomodels database: A repository of mathematical models of biological processes. *Methods in Molecular Biology*, *1021*, 189–199.

Chen, J., Densmore, D., Ham, T., Keasling, J., & Hillson, N. (2012). DeviceEditor visual biological CAD canvas. *Journal of Biological Engineering*, *6*(1).

Cooling, M. T., Rouilly, V., Misirli, G., Lawson, J., Yu, T., Hallinan, J., et al. (2010). Standard virtual biological parts: A repository of modular modeling components for synthetic biology. *Bioinformatics*, *26*, 925–931.

Courtot, M., Juty, N., Knüpfer, C., Waltemath, D., Zhukova, A., Dräger, A., et al. (2011). Controlled vocabularies and semantics in systems biology. *Molecular Systems Biology*, *7*, 543.

Eilbeck, K., Lewis, S., Mungall, C. J., Yandell, M., Stein, L., Durbin, R., et al. (2005). The sequence ontology: A tool for the unification of genome annotations. *Genome Biology*, *6*(R44).

Elowitz, M., & Leibler, S. (2000). A synthetic oscillatory network of transcriptional regulators. *Nature*, *403*, 335–338.

Endy, D. (2005). Foundations for engineering biology. *Nature*, *438*, 449–453.

Galdzicki, M., Peccoud, J., Wilson, M., Rodriguez, C. A., Oberortner, E., Pocock, M., et al. (2013). SBOL: A community standard for communicating designs in synthetic biology. http://dx.doi.org/10.6084/m9.figshare.762451.

Galdzicki, M., Rodriguez, C., Chandran, D., Sauro, H. M., & Gennari, J. H. (2011). Standard biological parts knowledgebase. *PLoS ONE*, *6*, e17005.

Galdzicki, M., Wilson, M., Rodriguez, C. A., Pocock, M. R., Oberortner, E., Adam, L., et al. (2012). Synthetic biology open language (SBOL) version 1.1.0. http://dx.doi.org/1721.1/73909 #87 (BBF RFC). http://openwetware.org/wiki/The_BioBricks_Foundation:RFC.

Goler, J. (2004). *BioJADE: A Design and Simulation Tool for Synthetic Biological Systems*. (Master's thesis). Massachusetts Institute of Technology.

Hallinan, J., Gilfellon, O., Misirli, G., Phillips, A., & Wipat, A. (2013). Tuning receiver characteristics in bacterial quorum communication: An evolutionary approach using standard virtual biological parts. *ACS Synthetic Biology*, in press.

Ham, T. S., Dmytriv, Z., Plahar, H., Chen, J., Hillson, N. J., & Keasling, J. D. (2012). Design, implementation and practice of JBEI-ICE: An open source biological part registry platform and tools. *Nucleic Acids Research*, *40*, e141.

Hill, A. D., Tomshine, J. R., Weeding, E. M. B., Sotiropoulos, V., & Kaznessis, Y. N. (2008). SynBioSS: The synthetic biology modeling suite. *Bioinformatics*, *24*, 2551–2553.

Hillson, N. J., Rosengarten, R. D., & Keasling, J. D. (2012). j5 DNA assembly design automation software. *ACS Synthetic Biology*, *1*, 14–21.

Hoops, S., Sahle, S., Gauges, R., Lee, C., Pahle, J., Simus, N., et al. (2006). COPASIa COmplex PAthway SImulator. *Bioinformatics*, *22*, 3067–3074.

Hucka, M., Finney, A., Sauro, H. M., Bolouri, H., Doyle, J. C., Kitano, H., et al. (2003). The Systems Biology Markup Language (SBML): A medium for representation and exchange of biochemical network models. *Bioinform.*, *19*, 524–531.

Huynh, L., Tsoukalas, A., Kppe, M., & Tagkopoulos, I. (2013). SBROME: A scalable optimization and module matching framework for automated biosystems design. *ACS Synthetic Biology*, *2*, 263–273.

Madsen, C., Myers, C. J., Patterson, T., Roehner, N., Stevens, J. T., & Winstead, C. (2012). Design and test of genetic circuits using iBioSim. *IEEE Design & Test of Computers*, *29*, 32–39.

Marchisio, M., & Stelling, J. (2008). Computational design of synthetic gene circuits with composable parts. *Bioinformatics*, *24*, 1903–1910.

Mirschel, S., Steinmetz, K., Rempel, M., Ginkel, M., & Gilles, E. D. (2009). ProMoT: Modular modeling for systems biology. *Bioinformatics*, *25*, 687–689.

Misirli, G., Hallinan, J. S., Yu, T., Lawson, J. R., Wimalaratne, S. M., Cooling, M. T., et al. (2011). Model annotation for synthetic biology: Automating model to nucleotide sequence conversion. *Bioinformatics*, *27*, 973–979.

Mutalik, V., Guimaraes, J. C., Cambray, G., Lam, C., Christoffersen, M. J., Mai, Q. A., et al. (2013). Precise and reliable gene expression via standard transcription and translation initiation elements. *Nature Methods*, *10*, 354–360.

Mutalik, V., Guimaraes, J. C., Cambray, G., Mai, Q. A., Christoffersen, M. J., Martin, L., et al. (2013). Quantitative estimation of activity and quality for collections of functional genetic elements. *Nature Methods*, *10*, 347–353.

Myers, C. J. (2009). *Engineering genetic circuits*. London: Chapman and Hall/CRC.

Myers, C. J., Barker, N., Jones, K., Kuwahara, H., Madsen, C., & Nguyen, N.-P. D. (2009). iBioSim: A tool for the analysis and design of genetic circuits. *Bioinformatics*, *25*(21): 2848–2849.

Myers, C., Barker, N., Kuwahara, H., Jones, K., Madsen, C., & Nguyen, N.-P. (2009). Genetic design automation. In *IEEE/ACM international conference on computer-aided design* (pp. 713–716).

Peccoud, J., Anderson, J. C., Chandran, D., Densmore, D., Galdzicki, M., Lux, M. W., et al. (2011). Essential information for synthetic DNA sequences. *Nature Biotechnology*, *29*, 22.

Pedersen, M., & Phillips, A. (2009). Towards programming languages for genetic engineering of living cells. *Journal of the Royal Society Interface*, *6*, S437–S450.

Quinn, J., Beal, J., Bhatia, S., Cai, P., Chen, J., Clancy, K. et al. ((2013). Synthetic biology open language visual (SBOL Visual) version 1.0.0. BioBricks Foundation Request For Comments (BBF RFC) #93 http://openwetware.org/wiki/The_BioBricks_Foundation:RFC.

Ro, D.-K., Paradise, E. M., Ouellet, M., Fisher, K. J., Newman, K. L., Ndungu, J. M., et al. (2006). Production of the antimalarial drug precursor artemisinic acid in engineered yeast. *Nature*, *440*, 940–943.

Stevens, J. T., & Myers, C. J. (2013). Dynamic modeling of cellular populations within ibiosim. *ACS Synthetic Biology*, *2*, 223–229.

Trankle, F., Zeitz, M., Ginkel, M., & Gilles, E. (2000). PROMOT: A modeling tool for chemical processes. Mathematical and Computer Modelling of Dynamical Systems, 6, 283–307. http://dx.doi.org/10.1076/1387-3954(200009)6:3;1- I;FT283.

Umesh, P., Naveen, F., Rao, C., & Nair, A. (2010). Programming languages for synthetic biology. *Systems and Synthetic Biology*, *4*, 265–269. http://dx.doi.org/10.1007/s11693-011-9070-y.

Xia, B., Bhatia, S., Bubenheim, B., Dadgar, M., Densmore, D., & Anderson, J. C. (2011). Developer's and user's guide to clotho v2.0: A software platform for the creation of synthetic biological systems. In C. Voigt (Ed.), *Synthetic biology, Part B computer aided design and DNA assembly* (pp. 97–135). *Methods in enzymology*. Vol. 498. San Diego: Academic Press Ch. 5.

Index

Note: Page numbers followed by *f* indicate figures and *t* indicate tables.

A
Arsenic biosensor, 77–79
Artificial neural networks (ANN), 23–26, 24*f*, 25*f*

B
BacilloBricks system, 109–111, 186, 186*f*
Bacillus functional analysis (BFA) mutant collection, 93, 95–96
Bacillus subtilis
  bet-hedging strategy, 88
  biosensors, 108–109
  computational tools and resources
    BacilloBricks system, 109–111
    DBTBS, 109–110
    population-level modelling, 110–111
    SubtiList database, 109
    SubtiWiki and SubtiPathways, 109–110
    transcriptomics, 109–110
  genome
    annotation, 89
    conjugation, 91
    engineering, 108
    gene dosage effect, 88–89
    gene transfer and recombination, 89–90
    minimalisation, 102
    size, 88
    transformation, 90–91
    vectors (*see* Plasmid vectors)
  growth medium, 87
  iGEM, 107–108
  International culture collection, 87
  metabolome analysis, 106–107
  proteome analysis, 105–106
  transcriptome analysis
    population analysis, 105
    reporter gene libraries/live cell arrays, 94*f*, 104
    reporter gene technology, 104
    sigma factors, 103–104
    single cell analysis, 105
    transcription and transcription profiling, 102
    transcription termination elements (TTEs), 104
BioBricks, 2, 39–40
BioFab, 184–185, 185*f*
BioJADE, 188, 188*f*
BioModels Database, 182, 182*f*
Biosensors
  advantages, 120
  application, 119
    bacterial SOS system, 121
    biomolecule properties, 121–122
    natural sensor mechanisms, 122
    regulatory elements, 121
  Bacillus subtilis, 108–109
  development, 152
  in vitro
    cell-free system, 141–142
    construction, 142–143
  limitations, 120
  modelling process
    biochemical model, 143
    constructing, 145–147, 146*f*
    guidelines, 145
    mathematical properties, 144
    mechanistic model, 143
    model-numerical simulation, 149–150, 150*t*, 151*f*
    model reduction, 147–148
    parameter values, 148–149, 149*t*
    phenomenological model, 144
    starting concentration, 148–149, 149*t*
    validation, 150–151
  reporters
    bioluminescence, 128
    colour change, 128–129
    fluorescence, 128
  research laboratory, 152–153
  synthetic biology, 129
  types
    different classes, 124, 124*t*
    posttranslational (*see* Posttranslational biosensors)
    protein synthesis stage, 122, 123*f*
    response characteristics, 124–128, 126*f*
    transcription-based (*see* Transcription-based biosensors)
    translation-based (*see* Translation-based biosensors)
Brute force techniques, 9

C
Campbell-type integration, 92–93
Cell casing
  cytoskeleton, 46–47

Cell casing (*Continued*)
  hypothesis, 46
  mur–fts clusters, 46
  osmotic pressure, 46
  protection, 43–44
  transport, 44–46
Chassis-related information transfer
  expression of program, 47
  macromolecules synthesis
    degradation, 52–53
    energy-dependent contraptions, 47–48
    energy-dependent helicases, 47–48
    entropy-driven processes, 47–48
    genes distribution, 52
    mould, 49–50
    proofread, 50
    protein biosynthetic machinery, 48
    protein shape, 48–49
    repair, 50–52, 51f
    scaffold, 48
    trash, 53
  regulation
    connect, 53
    control switches, 54
    epigenetic memory, 55
    genetic memory, 54–55
    speed control, 54
    stress gauge, 54
  replication, 47
Clotho, 197–198, 197f
Computational intelligence (CI)
  advantages and disadvantages, 27, 27t
  categories, 10
  ensemble approach, 10
  evolutionary algorithms (*see* Evolutionary algorithms (EA))
  fuzzy systems, 26
  neural networks, 23–26, 24f, 25f
  in silico evolution, 28
  swarm intelligence, 26–27
Cre/*lox* system, 98–100, 99f

D

Database of transcriptional regulation in *Bacillus subtilis* (DBTBS), 109–110
DeviceEditor, 195, 196f

E

Eugene, 197–198
Evolutionary algorithms (EA)
  advantages, 12–13
  annotation, 22–23
  circuit design categories, 14, 15t
  commonalities, 22
  design decision, 11
  evolutionary computation, 10–11, 11f
  features, 21–22
  fitness function, 11–12
  modularity
    crosstalk, 15–16
    Hill kinetics, 17
    interacting genes, 16
    logic gates, 17–18, 18f
    mass action kinetics, 17
    Michaelis–Menten kinetics, 17
    network motifs, 16–17, 17f
    posttranscriptional modification, 19
    in silico evolution, 18
  naïve algorithm, 13
  parent solutions, 12
  repressilator, 14
  simulated annealing, 20–21, 21f
Evolutionary computation (EC), 10–11, 11f
Expression vectors, 101

F

Fluorescence-activated cell sorting (FACS), 137, 138f, 159f, 161
Fluorescence in situ hybridization (FISH), 162
Förster resonance energy transfer (FRET) biosensors, 125, 126f, 128, 139, 142f
Fuzzy systems, 26

G

GenBank, 182–183, 183f
Gene expression
  heterogeneity
    noise-regulated (grey) virulence factors and traits, 165, 166t
    Pneumococcal pili (*see* Pneumococcal pili)
    Salmonella Typhimurium, 168–170, 169f
  noise (*see* Noise)
Genetic design automation (GDA)
  genetic circuit design workflow, 177–179, 178f
  repositories, 181
    BacilloBricks, 186, 186f
    BioFab, 184–185, 185f
    BioModels Database, 182, 182f
    GenBank, 182–183, 183f
    International Genetically Engineered Machine (iGEM) registry, 183, 184f
    Inventory of Composable Elements (ICE), 187, 187f
    standard biological parts knowledgebase, 184, 185f

SBML (*see* Systems Biology Markup Language (SBML))
SBOL (*see* Synthetic Biology Open Language (SBOL))
software tools, 187, 198, 199*t*
  BioJADE, 188, 188*f*
  Clotho, 197–198, 197*f*
  dataflow graph (DFG), 197
  DeviceEditor, 195, 196*f*
  Eugene, 197–198
  genetic engineering of cells (GEC), 192–193, 193*f*
  GenoCAD, 188–189, 189*f*
  Genome Compiler, 195–197
  Intelligent Biological Simulator (iBioSim), 194–195, 195*f*
  Kera, 194, 194*f*
  process modelling tool (ProMoT), 190, 191*f*
  Synthetic Biology Reusable Optimization Methodology (SBROME), 190–192, 193*f*
  Synthetic Biology Software Suite (SynBioSS), 190, 192*f*
  TinkerCell, 189, 190*f*
  VectorEditor, 195, 196*f*
Genetic engineering of cells (GEC), 192–193, 193*f*
GenoCAD, 188–189, 189*f*
Genome Compiler, 195–197

I

Intelligent Biological Simulator (iBioSim), 194–195, 195*f*
International Genetically Engineered Machines (iGEM), 77, 107–108, 121, 183
Inventory of Composable Elements (ICE), 187, 187*f*
In vitro biosensor
  cell-free system
    application, 141
    characteristics, 141–142
    construction, 142–143

K

Kera, 194, 194*f*

M

Macromolecules synthesis
  degradation, 52–53
  energy-dependent contraptions, 47–48
  energy-dependent helicases, 47–48
  entropy-driven processes, 47–48
  genes distribution, 52
  mould, 49–50
  proofread, 50
  protein biosynthetic machinery, 48
  protein shape, 48–49
  repair, 50–52, 51*f*
  scaffold, 48
  trash, 53
MatchMaker, 197–198

N

Noise
  engineering
    constructing robust gene circuits, 162–163
    feedback mechanism, 164
    finite number effect, 164
    gene expression, 159*f*, 163
    transcription fidelity, 164
    translation rate, 163
  heterogeneity, 164–165, 166*t*
  measurement
    FACS, 159*f*, 161
    FISH, 162
    gene expression, 159*f*, 160–161
    green fluorescent protein, 160–161
    RNA-fluorophore complex, 162
  origins
    extrinsic noise, 158
    gene expression, 158, 159*f*
    intrinsic noise, 158
    transcriptional bursts, 158–160

P

Plasmid DNA transformation, 90–91
Plasmid vectors
  antibiotic-resistance, 92
  expression vectors, 101
  integration
    BFA mutant collection, 93, 95–96
    clean deletion mutation, 96–97, 96*f*
    Cre/*lox* system, 98–100, 99*f*
    double-crossover recombination, 97, 98, 98*f*
    pMUTIN, 93, 94*f*
    single-crossover recombination, 92–93, 94*f*
    target genes, 93–94, 94*f*
    target insertion site, 97–98
    uses, 92
  pLOSS* vector, 100–101
  theta-replicating plasmids, 91–92
pMUTIN vectors, 93, 94*f*
Pneumococcal pili
  pili formation, 166–167, 167*f*
  vaccine, 167, 167*f*
  virulence factor, 165

Posttranslational biosensors
  application, 139
  characteristics of, 124t
  construct
    assembling, 140, 140f
    planning, 139–140
    testing, 141, 142f
  Förster resonance energy transfer, 139
Process modelling tool (ProMoT), 190, 191f
Proto BioCompiler, 197

R

Responsible research and innovation (RRI) approach, 81–82

S

Salmonella Typhimurium
  heterogeneity, 168, 169f, 170
  type 3 secretion system (T3SS), 168, 169f, 170
Sigma factors, 103–104
Simulated annealing, 20–21, 21f
Social dimensions
  governance, 80
  implications, 70
  metaphors and analogies, 71–72
  public acceptance
    arsenic biosensor, 77–79
    DIYbio movements, 76
    genetically modified crops, 76
    public good, 79
    technology safety, 75
  RRI approach, 81–82
  speculation and anticipation, 72–75
Standard biological parts knowledgebase (SBPkb), 184, 185f
Standard virtual parts (SVPs), 186, 186f
Synthetic Biology Open Language (SBOL), 198, 199t
  BacilloBricks, 186
  device and system element, 180–181
  DeviceEditor, 195, 196f
  GenBank, 182–183
  GenoCAD, 188–189
  iGEM registry, 183
  Intelligent Biological Simulator (iBioSim), 194–195
  interaction and participation element, 180–181
  Inventory of Composable Elements (ICE), 187, 187f
  Kera, 194
  Pigeon tool, 197–198
  Standard biological parts knowledgebase (SBPkb), 184, 185f
  TinkerCell, 189
  unified modelling language (UML) diagram, 180–181, 181f
Synthetic Biology Reusable Optimization Methodology (SBROME), 190–192, 193f
Synthetic Biology Software Suite (SynBioSS), 190, 192f
Systems Biology Markup Language (SBML), 198, 199t
  BacilloBricks, 186
  BioModels Database, 182
  development of, 179–180
  features of, 179–180, 180f
  genetic engineering of cells (GEC), 192
  GenoCAD, 188–189
  Intelligent Biological Simulator (iBioSim), 194–195
  Kera, 194
  process modelling tool (ProMoT), 190
  Synthetic Biology Software Suite (SynBioSS), 190
  TinkerCell, 189
Swarm intelligence, 26–27
Synthetic bacterial chassis
  bottom-up framing, 40, 41
  cell casing
    cytoskeleton, 46–47
    hypothesis, 46
    mur–fts clusters, 46
    osmotic pressure, 46
    protection, 43–44
    transport, 44–46
  clocks, 57
  exploration, 58
  helper function, 43
  information transfer (*see* Chassis-related information transfer)
  intermediate metabolism
    channel, 55
    inactivate, 55–56
    proofread, 56
    protection/unprotection, 56
    repair, 56
    salvage, 56–57
    trash, 56
  master functions, 43
  measuring time, 57
  time-dependent deformation, 57–58, 58f
  top-down framing
    definition, 40
    dendritic hierarchy, 41, 42
    fractals, 42

functional analysis, 40, 40f
'guilt by association' inference, 42
helper function, 40
human artefacts designing, 41
master function, 40
persistent genes, 41
scale-free structures, 42
segmented hierarchy, 42
Synthetic microbial genetic systems
  computational infrastructure
    BioBricks, 2
    bottom-up vs. top-down design, 3–5, 5f
    computer-aided design, 5–6
    design space concept, 8–9, 8f
    DNA synthesis and sequencing, 3
    standard virtual parts, 6–8, 8f
  computational intelligence (CI)
    advantages and disadvantages, 27, 27t
    categories, 10
    ensemble approach, 10
    evolutionary algorithms (see Evolutionary algorithms (EA))
    fuzzy systems, 26
    neural networks, 23–26, 24f, 25f
    in silico evolution, 28
    swarm intelligence, 26–27

large-scale designs, 2
small-scale designs, 1–2

T

Theta-replicating plasmids, 91–92
TinkerCell, 189, 190f
Transcription-based biosensors
  characteristics of, 124t
  construct
    assembling, 131–132
    planning, 129–131, 130f
    testing, 132–134, 133f
Transcription termination elements (TTEs), 104
Translation-based biosensors
  characteristics of, 124t
  construct
    planning, 135–136, 137f
    riboswitch construction guidelines, 136–138, 138f
  RNA switch, 134–135, 135f

U

Unified modelling language (UML), 180–181, 181f

V

VectorEditor, 195, 196f

PLATE 1 (Fig. 1.3 on page 8 of this volume).

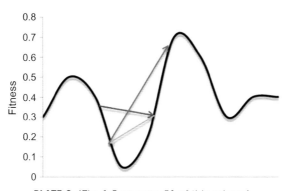

PLATE 2 (Fig. 1.8 on page 21 of this volume).

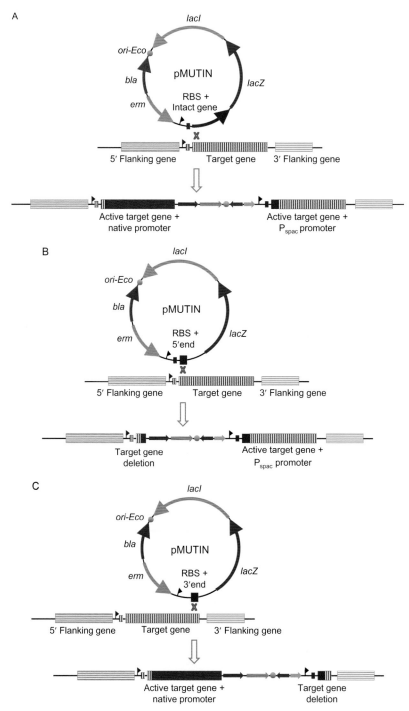

PLATE 3 (Fig. 4.1a–c on pages 94–95 of this volume).

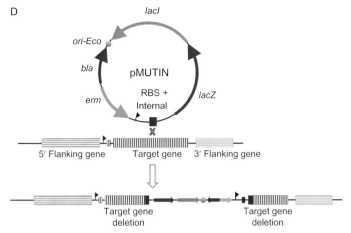

PLATE 3—Cont'd (Fig. 4.1d on page 95 of this volume).

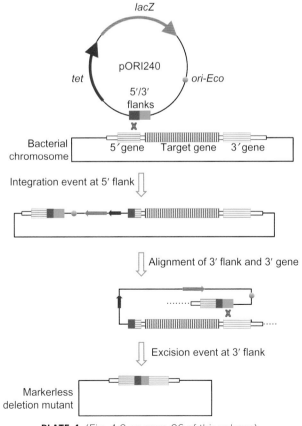

PLATE 4 (Fig. 4.2 on page 96 of this volume).

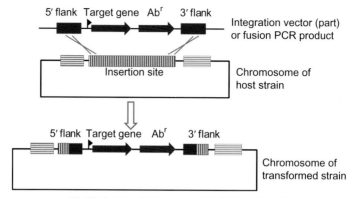

PLATE 5 (Fig. 4.3 on page 98 of this volume).

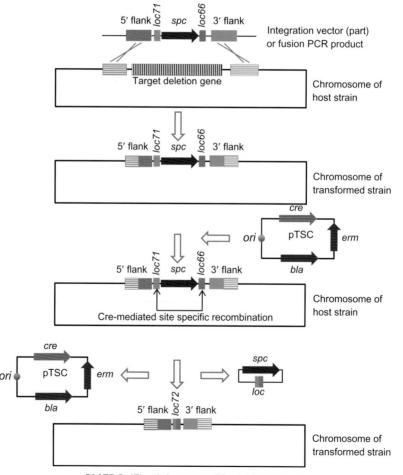

PLATE 6 (Fig. 4.4 on page 99 of this volume).

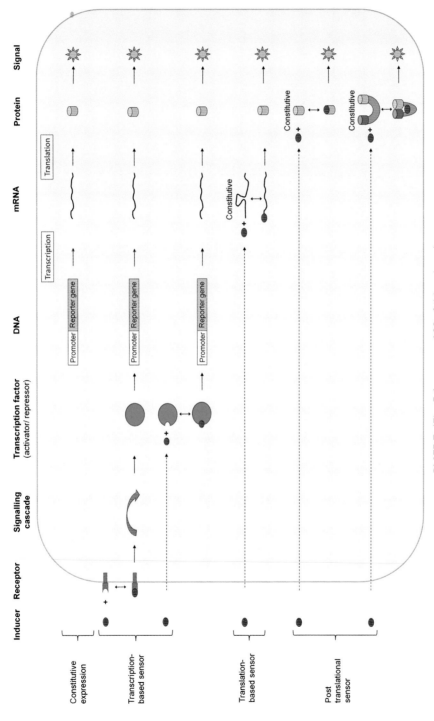

PLATE 7 (Fig. 5.1 on page 123 of this volume).

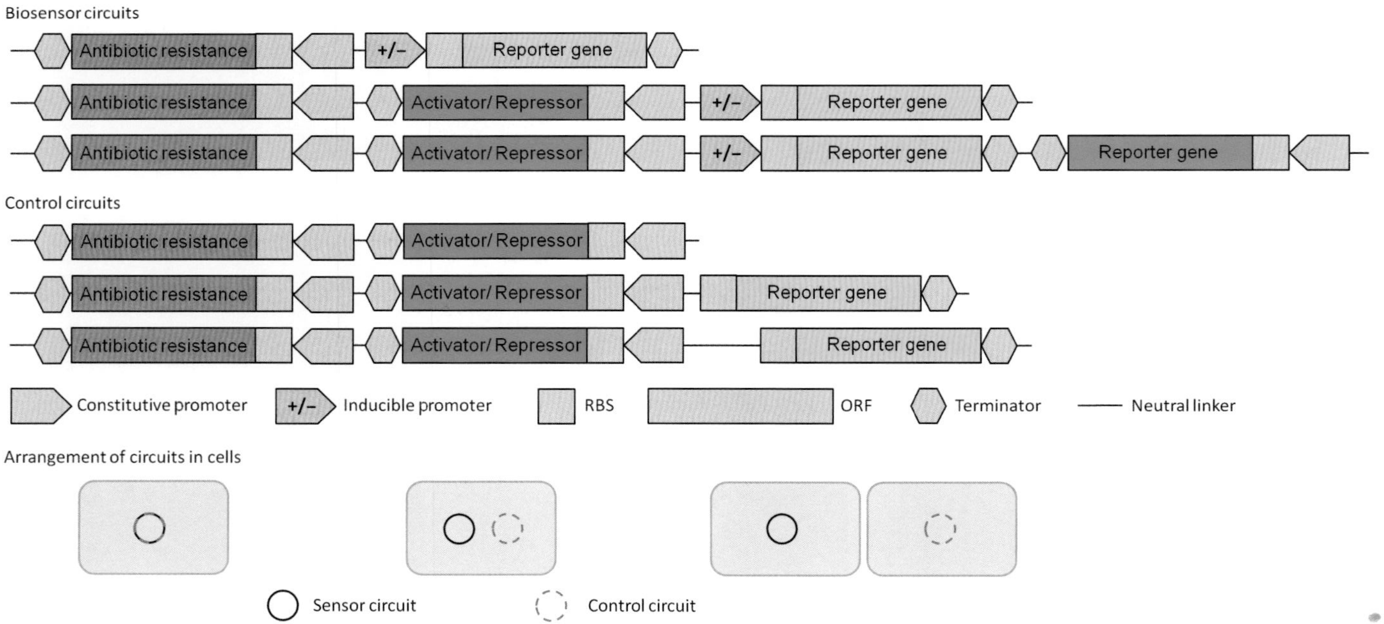

PLATE 8 (Fig. 5.4 on page 130 of this volume).

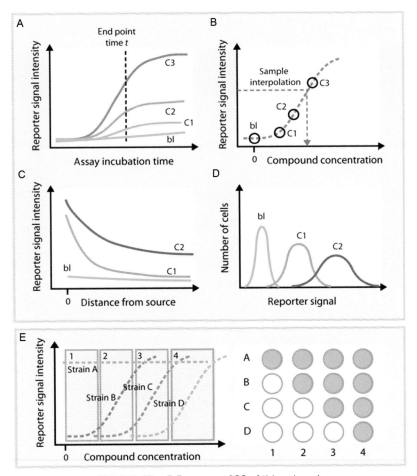

PLATE 9 (Fig. 5.5 on page 133 of this volume).

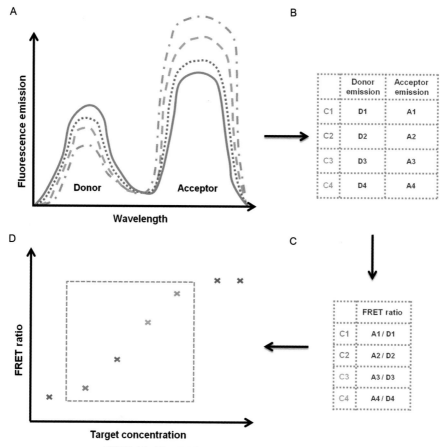

PLATE 10 (Fig. 5.10 on page 142 of this volume).

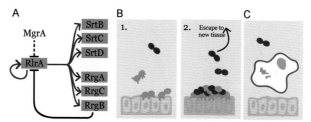

PLATE 11 (Fig. 6.2 on page 167 of this volume).

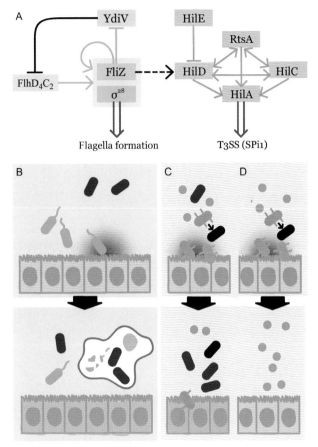

PLATE 12 (Fig. 6.3 on page 169 of this volume).

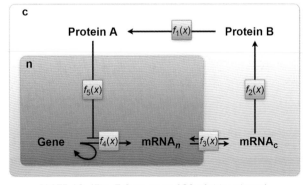

PLATE 13 (Fig. 7.2 on page 180 of this volume).

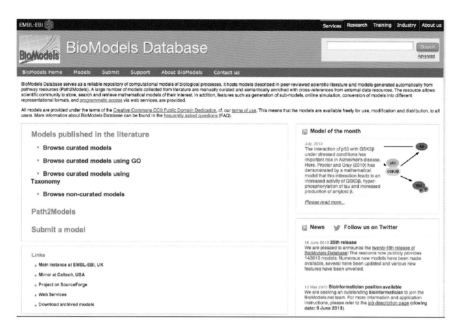

PLATE 14 (Fig. 7.4 on page 182 of this volume).

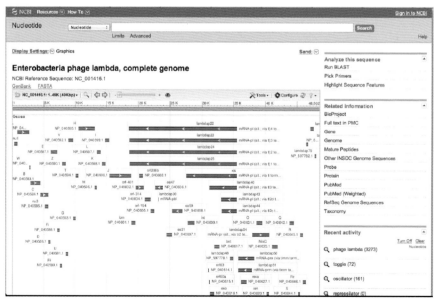

PLATE 15 (Fig. 7.5 on page 183 of this volume).

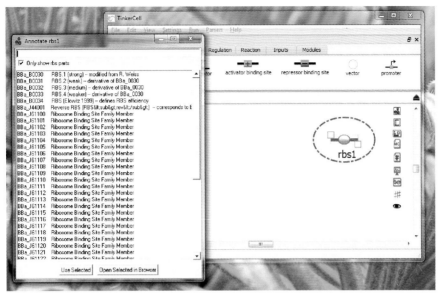

PLATE 16 (Fig. 7.7 on page 185 of this volume).

PLATE 17 (Fig. 7.9 on page 186 of this volume).

PLATE 18 (Fig. 7.10 on page 187 of this volume).

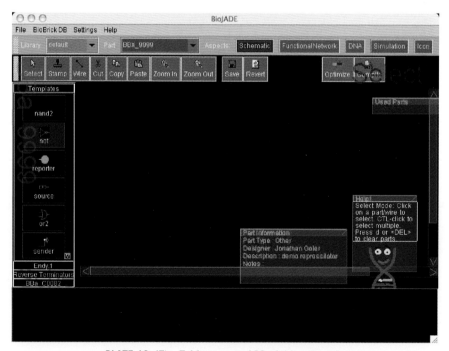

PLATE 19 (Fig. 7.11 on page 188 of this volume).

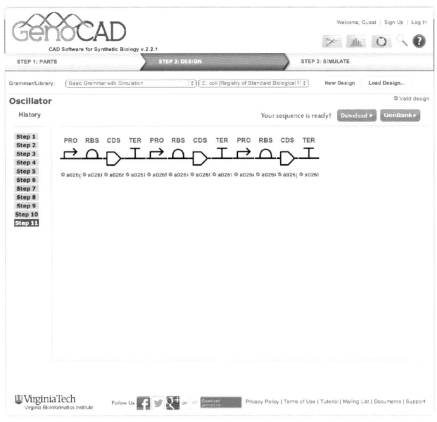

PLATE 20 (Fig. 7.12 on page 189 of this volume).

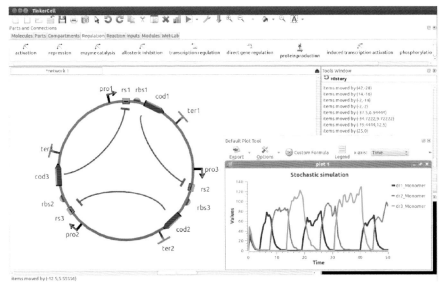

PLATE 21 (Fig. 7.13 on page 190 of this volume).

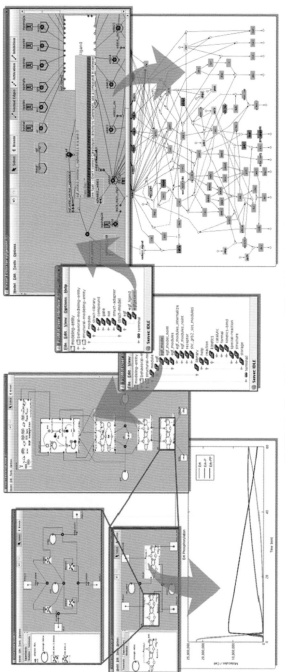

PLATE 22 (Fig. 7.14 on page 191 of this volume).

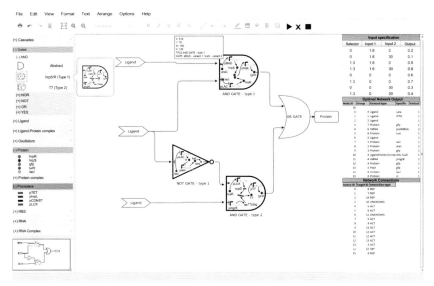

PLATE 23 (Fig. 7.16 on page 193 of this volume).

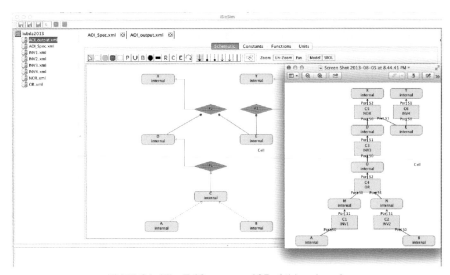

PLATE 24 (Fig. 7.19 on page 195 of this volume).

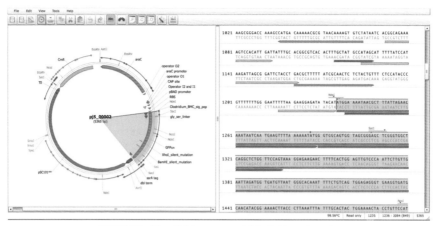

PLATE 25 (Fig. 7.21 on page 196 of this volume).

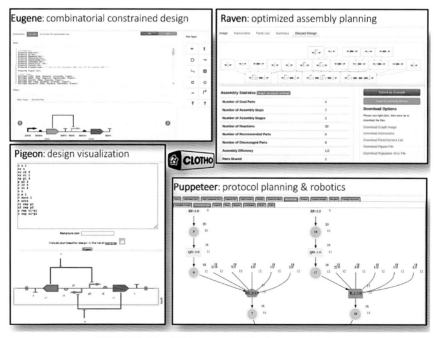

PLATE 26 (Fig. 7.22 on page 197 of this volume).